BIO TECHNOLOGY

AND THE
CHANGING ROLE OF
GOVERNMENT

ORGANISATION FOR ECONOMIC CO-OPERATION AND DEVELOPMENT

Pursuant to article 1 of the Convention signed in Paris on 14th December, 1960, and which came into force on 30th September, 1961, the Organisation for Economic Co-operation and Development (OECD) shall promote policies designed:

- to achieve the highest sustainable economic growth and employment and a rising standard of living in Member countries, while maintaining financial stability, and thus to contribute to the development of the world economy;
- to contribute to sound economic expansion in Member as well as non-member countries in the process of economic development; and
- to contribute to the expansion of world trade on a multilateral, non-discriminatory basis in accordance with international obligations.

The original Member countries of the OECD are Austria, Belgium, Canada, Denmark, France, the Federal Republic of Germany, Greece, Iceland, Ireland, Italy, Luxembourg, the Netherlands, Norway, Portugal, Spain, Sweden, Switzerland, Turkey, the United Kingdom and the United States. The following countries acceded subsequently through accession at the dates hereafter: Japan (28th April, 1964), Finland (28th January, 1969), Australia (7th June, 1971) and New Zealand (29th May, 1973).

The Socialist Federal Republic of Yugoslavia takes part in some of the work of the OECD (agreement of 28th October, 1961).

Publié en français sous le titre:

BIOTECHNOLOGIE :
ÉVOLUTION DU RÔLE DES POUVOIRS PUBLICS

This publication consists of several parts, presented in chronological order. It is suggested that the reader who wishes to begin with the most important texts, should first read the *Policy Recommendations of the Canada-OECD Joint Workshop on National Policies and Priorities in Biotechnology* (pp. ....), and continue with the *Opening Address of the Hon. Frank Oberle, P.C., M.P.*, Canadian Minister of State for Science and Technology.

Biotechnology and the Changing Role of Government is the fourth in a series of biotechnology studies which began in 1982, when the OECD published a scientific and technological state-of-the-art review (*Biotechnology — International Trends and Perspectives*, by A.T. Bull, G. Holt, M.D. Lilly) which has since received wide international recognition. After a discussion of that review, the OECD Committee for Scientific and Technological Policy (CSTP) decided to continue its biotechnology work, emphasizing four themes: patents, safety, R&D policies, and economic impacts. An expert report on patents appeared in 1985 (*Biotechnology and Patent Protection — An International Review*, by F.K. Beier, R.S. Crespi, J. Straus) followed in 1986 by a report on *Recombinant DNA Safety Considerations*, which included the Recommendation of the OECD Council concerning Safety Considerations for Applications of Recombinant DNA Organisms in Industry, Agriculture and the Environment.

The new report attempts to shed light on the current roles of government with regard to biotechnology, and to indicate where these are changing, thus allowing for a clearer separation between governmental and industrial roles. The first part, written by a consultant, Miss N. Field, reviews R&D policy issues and responses. The second presents the results of the Canada-OECD Joint Workshop on National Policies and Priorities in Biotechnology which took place in Toronto in April 1987. This workshop, organised jointly by Mr. David Shindler from the Ministry of State for Science and Technology, Ottawa, and Mr. Salomon Wald from the OECD, focused less on R&D than on issues and government roles in the diffusion of biotechnology.

Both parts are organically linked. They represent parallel efforts to address from different angles the question of national policies in biotechnology.

The views expressed in this report, including the recommendations of the Toronto workshop, do not necessarily represent the views of the OECD or of its Member countries.

Also Available

EVALUATION OF RESEARCH. A Selection of Current Practices (1987)
(92 87 05 1) ISBN 92-64-12981-2   78 pages   £5.00   US$11.00   F50.00   DM22.00

STI – SCIENCE, TECHNOLOGY AND INDUSTRY REVIEW (Half-yearly, Spring/ Autumn)
No. 2   September 1987
(90 87 01 1) ISBN 92-64-13002-0   174 pages

£8.00	US$16.00	F80.00	DM35.00

1988 Subscription

£15.00	US$30.00	F150.00	DM66.00

RECOMBINANT DNA SAFETY CONSIDERATIONS. Safety Considerations for Industrial, Agricultural and Environmental Applications of Organisms Derived by Recombinant DNA Techniques (1986)
(93 86 02 1) ISBN 92-64-12857-3   70 pages   £6.00   US$12.00   F60.00   DM27.00

BIOTECHNOLOGY AND PATENT PROTECTION. An International Review by F.K. Beier, R.S. Crespi et J. Straus (1985)
(93 85 05 1) ISBN 92-64-12757-7   134 pages   £8.00   US$16.00   F80.00   DM35.00

BIOTECHNOLOGY. International Trends and Perspectives by Alan T. Bull, Geoffrey Holt, Malcolm D. Lilly (1982)
(93 82 01 1) ISBN 92-64-12362-8   84 pages   £5.50   US$11.00   F55.00   DM28.00

CONTENTS

Part I

BIOTECHNOLOGY R&D:
NATIONAL POLICY ISSUES AND RESPONSES

by
Nancy Field

PREFACE

The following study, a consultant's report written for the OECD, was undertaken and completed much later than the preceding three OECD reports on biotechnology. However, any earlier comparative study on national R&D policies would have been outdated almost as soon as it was completed; institutional and policy changes affecting biotechnology have been occurring rapidly in many countries which have been intensely aware of the opportunities of this new technology. Only since 1985 have many OECD countries formulated a clearer and more comprehensive view of what they expect from biotechnology, or begun to establish a policy and institutional framework in which to discuss such views and act in accordance with them. In fact, an earlier OECD attempt to elicit government responses on national R&D policies in the field of biotechnology, undertaken by a questionnaire in 1983, failed completely because very few governments were in a position to respond at that time.

Investigation of national science and technology policies is a traditional topic for the OECD, and has always been on the agenda of the CSTP, unlike policies concerning intellectual property or safety and regulations which are quite new. In the case of the present study, however, more difficulties presented themselves than are normally encountered in science and technology policy studies. The number of agencies or ministries involved in biotechnology is greater than in most other technologies, and this has more than once led to conflicts in policy or in its interpretation. An additional and critical impediment has been the dearth of internationally comparable and comprehensive R&D statistics, based on an accepted and common definition of biotechnology — hence the deliberate exclusion of quantitative comparisons in this report. Biotechnology is not a field or discipline, but a widely applicable, enabling technology that cuts across a wide spectrum of activities, which makes the collection of data particularly difficult.

In view of these complexities, it seemed appropriate to base this new policy analysis on personal interviews in Member countries, supported, where available, by published material. Thus, Miss Nancy Field, who has drafted this report, reviewed during the course of 1986-87 the policies of 15 of OECD's 24 Member countries, as well as those of the EEC. The countries included are: Australia, Austria, Belgium, Canada, Denmark, Finland, France, Germany, Italy, Japan, the Netherlands, Sweden, Switzerland, the United Kingdom, and the United States. For Australia, Austria, Finland and Italy, this review was accomplished mainly through written communication; in the eleven other countries, oral interviews were conducted with experts in government ministries and agencies, as well as with industry.

Time constraints limited this review to 15 OECD Member countries. As the enquiry advanced, however, it became clear that most major problems or issues debated in one country, could to some extent be found in others as well. Thus, although not every OECD country with important biotechnology projects and policies has been included, the main policy issues and problems raised in the OECD area in the context of biotechnology R&D have been mentioned. In selecting countries for inclusion in the study, a particular effort was

9

made to look beyond the biotechnology "big powers" and to include a large number of small and medium-sized OECD countries. The following pages demonstrate that neither the challenges of biotechnology, nor the policy responses seem to separate small and big countries into two clearly distinct camps.

To make the comparisons for a balanced policy analysis, it was necessary to be selective in the discussion of events and decisions. From an individual country perspective, other events which were not mentioned might appear equally noteworthy. However, adding more facts to the text in order to make the description of countries more complete would have risked destroying a balance which was not always easy to achieve.

The preparation of this report was supported by a generous grant from the Government of the Netherlands. In addition, the Government of Japan helped finance a visit to that country by the consultant. Particular thanks are due to these two countries and to Professor D. Thomas who assisted in the translation of the report into French.

This paper was discussed by the Committee for Scientific and Technological Policy at its 47th Session in June 1987. However, it does not necessarily reflect the views of the OECD or of its Member governments. If the report is useful, thanks are due to the officials and experts in government and beyond — too numerous to be mentioned by name — who gave their time to assist the author in her endeavour.

SUMMARY AND CONCLUSIONS

This consultant's report reviews the policies of 15 of OECD's 24 Member Countries in the field of biotechnology R&D. It is based on interviews as well as on published material, and is the first comparative analysis of biotechnology R&D policies which addresses the situation of both large and small countries. The report has been structured according to four conventional science and technology policy themes, within which biotechnology has been examined. These are:

— Organisation of National R&D Programmes;
— National Objectives and R&D Priorities;
— Commercial Exploitation of University and Government Research; and
— Training and Mobilisation of R&D Manpower.

Within the context of these themes, several issues have emerged as central to biotechnology policy throughout the OECD. These issues are summarised below.

Coherence of R&D strategies and interagency co-ordination

Biotechnology has presented a number of critical and unprecedented policy challenges to OECD governments. One of the most pervasive technologies of the 20th century, biotechnology involves a very broad range of actors and agencies. In response to the vast scope of biotechnology, some governments, adhering to long-established political or administrative traditions, have maintained autonomous shaping and implementation of R&D policies within agencies. Many other governments, however, in order to achieve a coherent approach to policy, have appointed cross-sectoral committees to review developments and to recommend policies, or have organised closely integrated strategies which transcend intra-departmental boundaries and attempt to create an R&D synergy among a broad array of actors.

The objectives of interagency co-ordination range from maximisation of resources between agencies, and oversight of possible gaps in research, to closer alignment of research and education with the needs of industry, and consistency in regulatory policies and procedures. Some approaches to co-ordination have so far demonstrated greater efficacy than others. In any case, these initiatives, by facilitating cross-sectoral communication and enabling policy harmonization, represent in some cases a novel response to the issue of coherence in science policy, and signify a positive development in view of the need for vertical and lateral communication among the various actors involved in biotechnology.

Enhancement of industrial competitiveness: the need for an international perspective

Although several OECD governments have acted in response to the pervasiveness of biotechnology, industrial competitiveness represents the central and overriding concern of national strategies. In many countries, industrial policy in biotechnology appears to entail the pursuit of judiciously selected industrial targets, based on expected trends or competitive strengths; nevertheless, the increasing emphasis worldwide by both governments and industry on commercial exploitation of biotechnology creates the possible risk for governments of sinking resources into over-pursued target areas. It is therefore critical that countries maintain an international perspective on worldwide developments and relative strengths in biotechnology, in order to facilitate the exploitation of market niches and to ensure that the advantages of research expertise and local market strengths are balanced against global competition.

While some OECD countries have designated a rather diffuse grouping of applied industrial R&D topics as priority areas, or have been unable or unwilling to identify official priorities at all, several other countries have established a cohesive or thematically related grouping of industrial priority areas in biotechnology. In many of those countries with a coherent and strategic set of priorities, government policy has been notably influenced by a dynamic private sector. In some of the other countries, insufficient access to information regarding global developments has already proven a serious obstacle to judicious selection of priorities and targets for industrial development.

Implications of biotechnology for health, agriculture, energy and environment policies

Despite recognition in all OECD countries that the impacts of biotechnology would extend to many non-industrial sectors, some of the broader social and economic issues involving health care, agriculture and the environment have not been fully addressed and in some cases have been entirely neglected. These issues include adaptation of national public health systems to the impacts of biotechnology, the challenges of biotechnology to current agricultural policies, and utilisation of biotechnology to hasten and improve environmental clean-up.

In *health* care, government policies exist in the form of financial support for biomedical research as well as support of innovation and R&D in pharmaceutical development. In some cases, government support for pharmaceutical innovation appears to reflect a growing emphasis on prevention and diagnosis as opposed to therapeutics. Developments in biotechnology have raised several unprecedented issues (e.g. gene therapy protocols) and have introduced new dimensions to several perennial issues (e.g. orphan drugs for rare or Third World diseases). Although some studies have examined these issues in depth, little concerted action has been taken at government levels to address them.

In *agriculture*, the impacts of biotechnology will not be confined to improvements and modifications in traditional means of production; its effects will be felt in an increasing convergence of agriculture and industrial practice. Several countries have launched fairly comprehensive programmes to support or accelerate R&D developments in the agri-food domain. Most significantly, these programmes already reflect a growing involvement of government with farms and agro-industry, as competition in the agricultural sector becomes increasingly reliant on high technology. Biotechnology also impinges on issues of agricultural surplus, trade and employment, alternative land use, and ecological damage. However, not much attention has been directed at government levels towards reconciling the anticipated benefits of biotechnology with these issues.

In the *energy* sector, most countries regard biotechnological alternatives to fossil fuels as presently uneconomical, although a considerable amount of research activity is being supported. Such support generally reflects a desire to develop greater self-sufficiency in the long-term, particularly should the relative cost of current energy sources rise.

In the *environment* sector, where biotechnology has a vital role to play in monitoring and clean-up, the extent of research throughout the OECD is disappointingly small. Often, environment ministries maintain a defensive position vis-à-vis biotechnology, focusing on regulating against hypothetical environmental risks, rather than exploiting the many ways in which biotechnology might help them to monitor and clean-up the environment.

The lack of support and scarcity of efforts in the area of *risk assessment* research is particularly troublesome and deserving of attention, since the exploitation of many developments in biotechnology depends upon advances in this area. Part of the problem is due to the fact that, even when governments or funding bodies have actively promoted risk assessment research, talented researchers are difficult to obtain as scientists are generally not attracted to this apparently less prestigious area of biotechnology research.

Promoting commercial exploitation of university research/preserving fundamental research

In many countries, the need to improve industry/university links represents an old, but paramount theme in science policy. With the emergence of biotechnology, this theme has resurfaced with greater urgency, prompting several governments with long-standing weaknesses in this area to remove barriers and adopt new initiatives to facilitate the commercial exploitation of university and government research. Many of these initiatives adhere to traditional patterns, such as financial support for co-operative R&D projects, exchanges of research personnel, and targeted support for industrially relevant research in universities. Several other developments, however, including the establishment of R&D companies by governments and a growing awareness of intellectual property issues, depart from tradition and reflect the intensified concern with which governments treat the issue of commercial exploitation in the context of biotechnology. Although it is still difficult to assess the relative effectiveness of these initiatives, they have in some cases notably accelerated commercialisation or improved university/industry links and should be continued.

However, the proliferation of industry/university research agreements has raised concern regarding the possible implications for fundamental research. A tendency in some countries to bias support towards industrially relevant research in the universities at the expense of fundamental research, to allow industrial research support to gradually supplant government support, or to limit free dissemination of scientific knowledge, could in the long run undermine the fundamental research base on which future progress in biotechnology depends.

Training and mobilising R&D manpower/education and public acceptance

The two predominant features of manpower needs in biotechnology are their multidisciplinary nature and high qualification profiles. According to general consensus, such multidisciplinary training in biotechnology should complement, rather than supplant the training of specialists within the underpinning disciplines. The ultimate goal of such training should be to prepare individuals for successful interaction in research teams. In addition, the exposure to practical industrial problems as a part of the training process has also become recognised as an efficient means of facilitating rapid integration of R&D manpower into industrial biotechnology.

Attempting to transcend the hurdles which old educational traditions raise, such as the separation of biological and engineering faculties, most of the reviewed countries have responded to the demands of biotechnology by reorienting traditional training programmes in order to achieve greater interdisciplinarity. These adaptations range from the institution of special interdisciplinary bachelor's or master's degree programmes, to short courses at the post-graduate level for academics and industrialists. Most countries have also striven to increase contacts between universities and industry at the training level in biotechnology.

Notwithstanding these efforts, nearly all countries are experiencing manpower shortages in a number of areas necessary for underpinning industrial development in biotechnology. In areas where demand has been satisfied, some countries have been relatively successful in producing highly skilled manpower at home, while others have combined expertise developed at home with expertise from abroad. In any event, training opportunities which allow for multidisciplinarity and industrial exposure should be strengthened throughout the OECD, particularly in countries relatively less advanced in biotechnology R&D.

Furthermore, educational institutions must not only confront the need to train R&D manpower, but should also address the issue of general biotechnology education as it relates to public acceptance of the technology. Many OECD countries have already encountered undercurrents of public hostility towards biotechnology. A lack of public acceptance could cause considerable delays in the diffusion of many harmless and beneficial biotechnology products and processes. The efforts of government (as well as of industry) to contribute to greater public awareness and understanding of biotechnology will be necessary before continued public support of biotechnology may be ensured. An increase in the exposure of students at the elementary and high school levels to the fundamental and applied aspects of biological science may be an effective way of encouraging greater public acceptance of biotechnology.

I. ORGANISATION OF NATIONAL BIOTECHNOLOGY PROGRAMMES

Coherence of R&D strategies and interagency co-ordination

During the first half of the 1980s, many OECD governments found themselves faced with a number of critical policy challenges presented by the emergence of new developments in biotechnology. While partially manifest in questions of potential risk and the need to establish effective safety protocols, the public policy considerations have also been linked with exploitation of the social and economic benefits associated with biotechnology. In a climate of growing international competition, many governments have moved with a sense of urgency to accelerate the realisation of these benefits through programmes designed to harness and commercially exploit national research and development capabilities in biotechnology.

Although most of these governments have pinpointed biotechnology development as a national priority, the programmes and policy mechanisms which they have established to foster its evolution reveal notable differences in approach and organisation. Some of these differences reflect long-established characteristics of political philosophy and administrative tradition, such as the centralisation characteristic of the French programme, in which the government assumes an active, intervening role, or the policy of non-interference adopted by the United States and Switzerland. In several countries, however, government R&D policies in support of biotechnology have diverged from traditional paths, striving to respond to the two most critical aspects of biotechnology which have influenced the nature and organisation of these policies: the wide-ranging scope of biotechnology (its techniques being relevant to industry, health care, agriculture, energy and the environment), and the frequently close proximity of research to application.

These two elements of biotechnology have introduced new parameters and pose unprecedented challenges to the traditional organisation and procedures of national R&D policies. Although some governments have maintained autonomous shaping and implementation of R&D policies within individual agencies, others have responded to these new parameters by appointing cross-sectoral biotechnology committees to review developments and recommend policy, or by organising closely integrated strategies which transcend intra-departmental boundaries and attempt to create an R&D synergy among a broad array of actors. These strategies seek to achieve lateral co-ordination of research policy or research activities among relevant government agencies and/or to accomplish vertical integration of research and educational policy at the university level with development policies in sectors such as industry and agriculture.

Of the fifteen OECD countries surveyed, four (Germany, France, the Netherlands and the United Kingdom) have, with varying degrees of success, endeavoured to achieve vertical and/or lateral co-ordination of R&D policies and programmes in biotechnology. Five countries (Austria, Canada, Denmark, Italy and the United States), have established co-ordinating committees but have not yet achieved or sought to achieve the extent or degree of integration sought by those countries mentioned above. The remaining six countries (Australia, Belgium, Finland, Japan, Sweden and Switzerland) have not established co-ordinated planning of publicly supported biotechnology R&D programmes, although some have taken initial steps in this direction.

Of the fifteen countries reviewed, *Germany* developed the first organised government strategy for biotechnology. Having implemented several programmes to co-ordinate and strategically link research in universities, the Max-Planck institutes, Frauenhofer institutes and government research laboratories with industrial R&D priorities, the German strategy, established in 1972, has achieved a coherent and vertically integrated framework for industrial support. Although the focus of the biotechnology programme, centrally administered under the Federal Ministry for Research and Technology (BMFT), is primarily industrial, non-industrial areas have been receiving increasing support. Comprehensive lateral co-ordination mechanisms do not exist, as BMFT provides predominant support for all areas of applied research in biotechnology; however, some co-operative arrangements exist with other agencies, notably a bilateral group set up with the Ministry for Agriculture (BML) to determine common research priorities and to arrange for special funding in areas such as plant breeding.

The *French* Mobilisation Programme represents another highly centralised government programme in support of biotechnology, the overall strategy of which is to mobilise the R&D infrastructure by injecting seed money into scientific and industrial priority areas. The programme, organised under the aegis of the Ministry of Research and Technology closely co-ordinates research policy and planning among various government ministries (principally agriculture and industry), universities and research centres.

In the *Netherlands*, biotechnology was selected in 1982 as the first strategic sector to be supported by the Government under a new and elaborate scheme for the promotion of innovation-oriented research. The Innovation-Oriented Research Programme in biotechnology (IOPb), designed primarily to sharpen the relevance of government-funded biotechnology research to the development of Dutch industry, seeks to co-ordinate the resource allocation of its principal research funding agencies, ZWO and STW, with industrial priority areas in biotechnology identified by and for the Dutch manufacturing industry.

In addition, taking into account the broad reaches of biotechnology in the public domain, the Dutch programme also integrates into its organisational structure individual programme committees responsible for promoting the useful application of biotechnology in areas of agriculture, the environment and public health. Although its objective and the bulk of its funds are focused on the manufacturing industry, its subdivision into these three additional categories creates a mechanism for co-ordinating additional research areas which may or may not impinge on industrial development, but are nevertheless deemed to be in the national interest, e.g. waste treatment R&D. Research support within each category is managed by its own programme committee, which boasts wide representation while maintaining independence from its associated government ministry. Furthermore, all committees are linked by an advisory board which co-ordinates activities, seeks to identify common areas of interest, and promotes co-ordination of resources to further develop these areas.

In the Netherlands, traditionally close links between the environmental sector and both agriculture and industry have paved the way for viable projects currently planned through the IOPb programme committees, for joint research in developing waste management systems. Of those countries reviewed, the Netherlands is the only country to have adopted a structure of independent, but loosely associated committees responsible for nearly every sector of biotechnology, rather than the more common format of one all-embracing committee chosen by other countries[1].

In the *United Kingdom*, the need for a cohesive R&D strategy in biotechnology was articulated in the recommendations of the "Spinks Report" (1980). This report was compiled by a Working Party for Biotechnology, under the auspices of the Advisory Council for Applied Research and Development, the Advisory Board of the Research Councils and the Royal Society, with a mandate to review prospects and to recommend action to facilitate British industrial development in biotechnology. Contending that biotechnology "straddles the divisions of responsibility both among Government Departments and among Research Councils" and would point up gaps between them, the Spinks Report recommended that the "activities of Government Departments in relation to biotechnology should be co-ordinated; (and) a coherent programme of industrial research and development, involving industry, government research establishments, universities and research councils should be pursued....".

This objective was achieved by forming an Interdepartmental Committee on Biotechnology (ICBT), with the UK Department of Trade and Industry as the lead Agency. The role intended for the ICBT has been to co-ordinate activities sponsored by individual Departments, and to provide a focus within Government for discussion and action. Although the ICBT has effectively served as a forum for discussion, and amongst other things has co-ordinated the UK response to research proposals from the European Commission, it has no funds of its own and has not exerted the degree of pro-active influence recommended by Spinks.

Dynamic co-ordination of biotechnology R&D has been achieved between the Biotechnology Unit of the UK Department of Trade and Industry (current expenditures approximately £7 million per annum) and the Biotechnology Directorate of the Science and Engineering Research Council (current expenditures approximately £4 million per annum). These two units establish priorities and programmes in close consultation with each other; as a result, industrial priorities are articulated to, and pursued by the research community, and vertical co-ordination is hence achieved. With regard to the remaining principal actors supporting biotechnology R&D, such as the Medical Research Council, the Agricultural and Food Research Council, and the Natural Environment Research Council, no precedent has been established for active lateral co-operation.

The *Canadian* biotechnology strategy, adopted in 1983, has been organised and administered under two committees, with the National Research Council (NRC) acting as the lead agency for R&D. The National Biotechnology Advisory Committee, with representation from industry, academia and government, advises the Minister of State for Science and Technology (MOSST) on development issues. An Interdepartmental Committee on Biotechnology, chaired by a representative of MOSST, co-ordinates federal government activity in biotechnology and has responsibility for the allocation of National Strategy Funds. Despite a reasonably comprehensive plan for biotechnology, however, neither the National Advisory Committee nor the Interdepartmental Committee has sufficient mandate or resources to assume an effective "pro-active" role, particularly given the tremendous strides which will be necessary before Canada can fully reap the opportunities presented by biotechnology[2]. Although under the Canadian strategy, priority areas for research have been identified and promoted, based on perceived opportunities to strengthen Canada's natural resource industries through biotechnology, funding appears to have been inadequate to build a coherent strategy based on those priorities, to support research programmes at universities and government institutes in co-ordination with industrial priority areas, and to ensure industrial involvement in these research programmes.

Furthermore, in contrast with the UK, Canada currently has no counterpart to the close interaction between the Biotechnology Unit of the Department of Trade and Industry (BTU) and the Science and Engineering Research Council (SERC) to build a foundation of industrially relevant research and protocols for industrial involvement in that research. In order to achieve greater strategic co-ordination between industrial development and university and government research, and to achieve long-term policy objectives, according to the Canadian Science Council, conflicting aims within and between federal and provincial agencies must be reconciled, and roles and responsibilities vis-à-vis biotechnology must be more clearly defined[3].

The Government strategy might also consider rectifying what has been termed a "fragmented and incoherent" research effort at universities and research institutes in biotechnology, by providing the Natural Sciences and Engineering Research Council (NSERC) with greater resources to support strategic projects, based on industrial or other priority areas[4].

In *Sweden*, several government agencies together with some industrial actors instituted a National Committee for Biotechnology under the auspices of The National Swedish Board for Technical Development. This Committee was mandated to stimulate co-operation between different bodies supporting R&D in biotechnology with the aim of co-ordinating research at the national level. However, upon completion of a national research programme, submitted to the government in 1986, the activities of the Committee were discontinued.

The *United States*' "Biotechnology Science Co-ordinating Committee", established in 1985, serves primarily to promote consistency amongst federal agencies in the development of scientific review procedures and assessments, rather than to co-ordinate the planning of R&D activities between agencies in order to pool resources or avoid overlaps. Membership on the committee includes The Department of Agriculture, The Food and Drug Administration, the National Institutes of Health, the Environmental Protection Agency and the National Science Foundation.

Neither *Australia* nor *Japan* has established systematic means for co-ordinating research and development policies in biotechnology. In *Belgium*, agencies have acted independently in formulating domestic R&D policies in biotechnology. However, recently efforts have been made in the context of science policy, through a "Group for Concertation in International Co-operation", to promote interdepartmental policy co-ordination regarding international initiatives, notably in biotechnology. In *Finland*, concerned branches of government are presently acting independently of each other in formulating and enacting R&D policies, but negotiations have begun between the administrative branches of the Ministries of Education and Trade and Industry for co-ordination of training and R&D in biotechnology. *Switzerland* has maintained a decentralised R&D policy in all sectors including that of biotechnology.

In *Italy*, a National Biotechnology Committee has been appointed to assess the opportunities presented by biotechnology for Italian industry (particularly pharmaceutical) and agriculture. According to recommendations published in July, 1986,[5] the committee advised the formation of an interdepartmental committee for biotechnology with representation from ministries of health, the environment, agriculture and from the public research establishment, based on the prototype of either the Netherlands or the UK. Furthermore, it suggested the long-term possibility of creating a new agency to handle co-ordination functions in biotechnology. The need for closer co-ordination between industrial priorities and areas of research supported heavily by the research establishment has been identified as critical for the strategic development of competitive strength in Italian biotechnology.

*
**

Given the recent development of most of these biotechnology programmes, it appears somewhat premature to draw firm conclusions regarding their overall effectiveness, although relative strengths and weaknesses are already apparent. One can safely say at this point, however, that the need for vertical and lateral communication at the policy level between the various actors involved in biotechnology is critical, whether the objective of such cross-sectoral communication be maximisation of resources between agencies, oversight of possible gaps in research, closer alignment of research and education with the needs of industry, or consistency in regulatory policies and procedures. The plethora of unprecedented initiatives in this direction in OECD countries during the past few years attests to growing recognition of this fact.

II. NATIONAL OBJECTIVES AND R&D PRIORITIES

1. The enhancement of industrial competitiveness

Although several OECD governments have acted in response to the pervasiveness of biotechnology, industrial competitiveness appears to represent the central and overriding concern of national strategies. This may be largely attributable to the fact that the recognition by governments of the opportunities presented by biotechnology for economic growth occurred at a time when one of their pre-eminent concerns was to seek and promote new ways to overcome economic recession. In several cases, direct pressure from industrial lobbies also played a significant role in influencing the industrial orientation of biotechnology policies.

The increasing emphasis worldwide by both governments and industry on commercial exploitation of biotechnology creates the possible pitfall for governments of sinking resources into over-pursued target areas where, for most of them, the chances for a return on investment will be negligible. It is therefore critical that countries maintain an international perspective on worldwide developments and relative strengths in biotechnology, in order to identify underexploited market niches, and to balance the advantages of research expertise and local market strengths in given areas against global competition.

a) Commercial opportunities and constraints

Since the initial breakthroughs in genetic engineering techniques in the early 1970s, and the proliferation of US biotechnology start-up companies in the pharmaceutical sector, biotechnology R&D has been undertaken to varying degrees within a much broader spectrum of industries[6]. It has penetrated the established pharmaceutical, chemical, food processing, pulp and paper and scientific and industrial equipment industries, and spawned a new generation of science-based companies already beginning to exploit biotechnology in the agricultural sector.

In the food-processing and chemical industries, the vast range of potential opportunities which biotechnology presents extends from replacing traditional chemical production processes with biological ones, e.g. the utilisation of genetically engineered organisms to produce chemical compounds such as alcohol, enzymes, amino acids, vitamins and oils, to improving biological fermentation processes through immobilised enzyme techniques, more efficient bioconversion processes based on modified organisms or more sophisticated monitoring and control systems, or finally, improved separation and purification techniques. Biotechnology will make possible the synthesis of novel compounds, the production of greater and purer quantities of biologics, new microbiological techniques for mineral leaching, oil extraction and waste treatment, alternative energy sources and new processes for the production of raw material feedstocks. In the agricultural sector, small companies are already exploiting niches in specialised agricultural and horticultural products designed for growers, processors and consumers, based on the use of protoplast fusion techniques, and products such as bovine growth hormone, pseudo-rabies vaccine and ice-minus bacteria are seeking commercialisation pending resolution of various regulatory and other issues.

Notwithstanding the tremendous technical potential of biotechnology to develop novel products and processes in a wide spectrum of industries, and to improve existing products and processes, the magnitude of the impact has been considerably less than originally anticipated, and market forecasts for biotechnology products vary markedly[7]. The reasons for this slower than anticipated pace and consequent uncertainty regarding the extent of biotechnology's impact may be attributed to a broad range of technical, commercial and policy-related factors. Unforeseen or underestimated technical barriers include insufficient knowledge of the relationship between the structure and the function of proteins, and the behaviour of genetically engineered micro-organisms under industrial conditions[8]. Factors in the policy arena extend from lengthy administrative and regulatory procedures before a product may be considered safe for field-testing or marketing, to the deliberate restriction of certain controversial products such as factory-produced aspartame in the EEC. Commercial barriers centre around the absence of viable markets for certain biotechnology products, e.g. biomass based fuel.

Because these factors, compounded by the physical constraints on biotechnology development (inadequate supplies of manpower, obsolete equipment and lack of data bases), introduce substantial financial risk and might discourage or prevent industrial R&D investment, many governments have implemented various policies to reduce the financial risk, to lessen the physical constraints, and to increase the incentives to conduct R&D in this field, precisely because despite all of the bottlenecks, the recognition exists that the ultimate payoffs of biotechnology, both economically and socially, could be enormous. Several countries have also identified national priority areas for development in order to optimise scarce resources. These measures have all been adopted in order to create a climate conducive to future competitive strength in the scientific and commercial development of biotechnology.

b) Government targets and priorities

Amongst the fifteen countries reviewed, five (Germany, Japan, France, Sweden and the Netherlands) have established and are supporting a cohesive or thematically related grouping of industrial priority areas; three countries (the United Kingdom, Austria and Australia) have designated a more diffuse grouping of applied R&D topics as priorities, yet (Austria excepted) with an evident emphasis on a select few areas; five countries (the United States, Switzerland, Finland, Belgium and Denmark) have, for a variety of reasons ranging from political philosophy to conflict or uncertainty, been unable or unwilling to identify official priorities for industrial biotechnology R&D; finally, in two countries (Canada and Italy) attention has been directed towards concentrating central government support on industrial priorities in biotechnology, but prevailing obstacles need to be surmounted before these objectives may be realised.

In *Germany*, shortly before the major scientific breakthroughs in genetic engineering occurred, the first organised government strategy in support of biotechnology was established in 1972 by the German Federal Ministry for Research and Technology (BMFT). In response to industrial lobbying concerning the growing importance of developments in bioprocess technologies[9] for the chemical and pharmaceutical industries, the government established a special programme to support R&D in that area. Since then, the BMFT's priorities for biotechnology development have evolved (particularly since 1982) to clearly underline advancements in cell culture, cell fusion and genetic engineering, as well as in bioprocess technology.

The current BMFT biotechnology programme remains primarily geared towards the development of advanced genetic and bioprocess technologies, and towards the large-scale introduction of these technologies in industry. The programme also seeks to promote the establishment of technology-based firms, and to lend support to small and medium-sized firms implementing new biotechnological techniques. The programme places a heavy emphasis on effectively linking fundamental and applied research expertise in Germany with industrial activities by means of incentive programmes for joint research in priority areas, and benefits from a traditionally close-knit relationship between the German research community, government establishments and industry.

The Government has highlighted gene technology through the BMFT's identification of four "gene centres", centres of excellence in genetics, pinpointed for special support and targeted as nuclei for joint project research with universities and private firms. In addition, government officials expect that centres for downstream processing and immunology will be identified in the near future. Biotechnology represents the first technology for which the BMFT has identified and supported research "centres" in this manner.

The German Research Society (DFG), which funds German basic research, has also contributed to developments in biotechnology. Amongst its various funding priorities are, notably, two concerted programmes in genetic engineering and bioprocess engineering. In addition, research involving various aspects of process technology, e.g., enzyme and cell culture technology, are prioritised by the GBF, Gesellschaft für Biotechnologische Forschung. A research establishment founded in 1968 by the Volkswagen Foundation to exploit developments in process engineering, the GBF is presently financed by the German Government, currently receiving 90 per cent of its funding from the BMFT.

The German biotechnology programme continues to give the preponderance of its support to complementary developments in bioprocess technology and genetic engineering; however, the proportion of federal support allocated to areas such as biosafety, protein synthesis, bioelectronics, plant breeding and renewable resources appears to be growing. Finally, in addition to federal government support, several of the federal Länder (states) support biotechnological research — sometimes through extensive special programmes — thereby complementing the R&D support mechanisms of the federal government. Most of these activities, however, concern the promotion of basic research in universities.

The *Japanese* industrial strategy in biotechnology, organised and administered under the auspices of the Ministry for International Trade and Industry (MITI), and based upon subsidies for "priority" research conducted by industrial consortia, appears equally focused, if not more foresighted and enterprising than that of Germany. In addition to a concerted effort to develop a belated competence in genetic engineering techniques, along with concurrent efforts to keep pace with improvements in process technologies (where Japan has outstanding traditional know-how), in order to apply and integrate them into Japan's chemical and fermentation industries, MITI has been pursuing many goals with less obvious, and probably much more distant market potential.

Although many governments are supporting research on biomass as a source of renewable energy, Japan is the first and so far the only OECD country amongst those surveyed with a massive government programme to develop this area, amounting to a Y 1 247 million allocation in 1985, and Y 1 312 million in 1986. Japan's total dependence on imported energy is clearly an added incentive to develop biomass research as an alternative energy source and a means of greater self-sufficiency, despite negative speculation about its potential market value due to currently low prices of oil (1985-86) and enormous development costs.

Waste water treatment and utilisation represents another massive government programme sponsored by MITI, the funding of which rose dramatically from Y 20 million in

1985, to Y 1 072 million in 1986. MITI's concentrated efforts to develop a waste water purification process based on biotechnology reflect a keen expectation amongst experts worldwide that waste treatment will represent one of the areas most profoundly affected by biotechnology developments.

MITI is also indirectly supporting the development of the Protein Engineering Research Institute which was established in 1986 and receives 70 per cent funding from the Japanese Key Technology Centre, largely funded by MITI, and the remaining 30 per cent from a consortium of 14 companies. Estimated costs of the centre are in the order of Y 17 billion covering construction and 10 years' operating costs. Finally, it deserves mention that the Japanese Government has recently included a programme on biochips in its "Basic Technologies for Future Industries" programme. Although this fledgling programme does not merit special attention for the relatively modest funding thus far allocated, its inclusion under the "Basic Technologies" programme may indicate the perception that biochips will become as widely applicable and as much an underpinning technology as rDNA, bioreactor development and large scale cell cultivation.

The *French* Government, under the aegis of the Programme Mobilisateur, has designated two prime target areas for priority support in biotechnology. These are the development of new diagnostics and vaccines, notably including a large project to develop an AIDS vaccine, as opposed to the more "classical" development of therapeutic drugs, and the development of greater expertise in the agri-food area. This second priority area, represented by a coherent programme entitled "Aliment 2000" covers such topics as improved control of the fermentation process for wine production, and enzyme utilisation in complex situations, e.g. involving low water content. Finally, although fundamental research efforts in France are admittedly weak in this area, the Programme Mobilisateur strongly supports industrial R&D on waste treatment processes, where two French companies (Companie Générale des Eaux, and Société Lyonnaise des Eaux) have particular strengths. In all cases, however, the Programme Mobilisateur only provides seed money to catalyse efforts in priority areas.

In *Sweden*, biotechnology was first defined as a priority area by the National Swedish Board for Technical Development (STU) in 1978, but it was not until 1980 that biotechnology began to receive priority support, allocations increasing at a rate of 10-20 per cent per annum, as STU's overall budget rose at a rate of approximately 4 per cent per annum. Industrial biotechnology policy in Sweden, under the auspices of the STU, has taken a conservative and focused approach to biotechnology development, concentrating resources on areas of research and development where Sweden already has an obvious competitive advantage, such as separation and purification technology. STU has identified the following priority areas, in descending order of importance, for industrial biotechnology: gene technology, downstream processing (purification methods), enzyme technology and microbial growth (entailing bioreactor design, scaling-up, computer monitoring and control, etc.) In fiscal year 1985-86, these areas received funding of approximately Skr 8 million, 3 million, 3 million and 3 million respectively.

STU is also supporting projects on bio-organic synthesis, biosensors and protein engineering. Finally, STU supports strategic areas of basic research with long-term potential for industrial use, generally corresponding to the aforementioned industrial priority areas. These are: molecular and cell biology, large-scale cell culture, downstream processing, and separation and purification techniques. Sweden has deviated very little from traditional directions of resource allocation, utilising the keen present interest in biotechnology to strengthen existing and to develop new areas of work in life sciences.

In the *Netherlands*, four categories of "applied research" topics have been identified as eligible for support under the Governments's Innovation Oriented Strategy for Biotechnology (IOPb). These are: host-vector systems, somatic cell hybridisation, second generation bioreactors and downstream processing. Projects in these areas may be eligible to receive 35 per cent of the IOPb's budget, or Gld 25 million. Other areas identified for support as priority programmes of the Advisory Committee Board, are protein engineering, soil biodegradation, plant biotechnology, and monoclonal antibodies, although for these targets, as compared with the aforementioned, significantly fewer funds are available. A further stimulus to R&D activities in biotechnology will be provided by a new government programme entitled the "Programmatic Company-Oriented Technology Stimulation Plan for 1987". According to this plan, R&D in four key technologies including biotechnology will be supported by means of financial stimuli, training and publication measures.

The strategy for industrial development in biotechnology in the *United Kingdom* is less focused and funding more dispersed than that of either Germany or Japan. The first set of priorities were articulated by the Department of Trade and Industry (DTI) in 1982, when, despite an initially negative response to Spinks' recommendations for an industrial biotechnology strategy inspired by anti-interventionist sentiment, DTI allocated a package of £16 million spread over a three year period to three designated "centres" of chemical engineering. The selection of the centres in areas of downstream processing, fermentation technology and second generation bioreactors indicates the priority status accorded to chemical engineering and advanced bioprocess technologies, due to an acknowledged weakness in those areas in the UK and recognition of their fundamental importance to industrial exploitation of biotechnology. The need to overcome weaknesses in these areas was perceived as particularly acute in light of UK's recurrent failures to translate its scientific and technical breakthroughs into commercial products and processes, and a growing frustration over the continual exploitation abroad of UK research expertise.

In 1982 as well, a Biotechnology Unit (BTU) in the Department of Trade and Industry, and a Biotechnology Directorate in the Science and Engineering Research Council were established. Both the BTU, staffed by secondees from industry and civil servants, and the SERC, in consultation with each other and with industry, have identified further "strategic" areas of support for industry. BTU has demonstrated some consistency with the Department's earlier initiatives by establishing a 5-year programme on second generation bioreactors, operated in conjunction with seven firms at the Institute for Biotechnological Studies (IBS), which has been recognised as a centre of excellence in this area[10]. The total cost of the programme has been estimated at £1.4 million pounds, of which BTU is supporting one half.

The Science and Engineering Research Council (SERC) has similarly taken initiatives to support bioprocess technology, through the establishment of a research "club" in downstream processing, a scheme whereby industrial funds are drawn to the support of university research in this area. Furthermore SERC has established a 'club' in protein engineering, involving Glaxo, ICI, Celltech and Sturge as participants, and consisting of a series of research programmes at UK universities; a co-operative venture in yeast research involving five firms has been established at the Biocentre at the University of Leicester.

In addition, SERC is independently pursuing and giving deliberate priority status to programmes with "foreseeable industrial applicability", e.g. biosensors and microbial physiology, but currently no industrial participants are involved in these programmes, even in a limited capacity. Other SERC priority areas in biotechnology, in descending order of importance based on funding levels, are rDNA technology, large scale growth of plant and animal cells, plant genetics and biochemistry, waste treatment and biodegradation, biocatalysis, and fermentation technology.

One of the most significant initiatives which the Biotechnology Unit of the Department of Trade and Industry has undertaken, has been the launching of a Collaborative Research Initiative in the Agri-Food sector, indicating a fairly recent decision to highlight the opportunities which industrial development in this sector represents for the UK. The agri-food sector does in fact represent a strategic area for concerted R&D investment in the UK, due both to the importance of this sector in international markets, the competitive quality of agricultural research in the AFRC institutes and agricultural faculties, and the vast anticipated potential for exploitation of biotechnology in this sector. The realisation of this potential, however, hinges on significant advances in fundamental research, for a greater understanding of biological processes will be essential before such processes may be modified for commercial purposes. In order to stimulate this 'low-tech' industry with an historically weak commitment to long-term R&D to adopt 'high-tech' approaches to production, the Government has assumed a role to provide greater incentives for R&D investment in this area, and to encourage better exploitation by industry of existing research expertise through closer links with university and government laboratories.

The outcome of this initiative is yet uncertain; one major programme has been established which draws together some 75 per cent of the UK research expertise in plant molecular genetics at the Plant Breeding Institute, the John Innes Institute and the Universities of Durham and Warwick; eleven companies share with DTI the cost of the £3M and 3-year programme.

Finally, the BTU has identified the following as additional priority areas eligible for extra support: Process Plant and Apparatus, where currently the UK is relatively quite weak and dependent on imports, Enzyme Use and Production, another weak area which has not been sufficiently (if at all) addressed, Biosensors, which has been identified as an area of great long-term strategic importance, and Diagnostics, where UK research is extremely competitive.

Priority areas in *Australia* for industrial and agricultural development in biotechnology are widely based on agricultural production, and human and veterinary health products, where strong research and some industrial base exist. Rather than targeting a select group of industrially relevant research areas for priority support, however, the Australian development strategy for biotechnology appears to place greater emphasis on improving the R&D investment climate through a variety of financing and fiscal measures available to a wide range of R&D activities including those relevant to biotechnology. The Australian Government has, nevertheless, cited as priorities for biotechnology R&D a rather broad and diverse grouping of areas, including plant agriculture, e.g. genetic engineering to modify important microbes, and tissue culture to improve plant species; animal production, e.g. diagnostic probes, vaccines, advanced breeding techniques, and animal hormones; human pharmaceuticals and health; equipment and instrumentation; food processing, and water treatment.

Canada's federal strategy for industrial biotechnology has identified as R&D priorities nitrogen fixation, cellulose utilisation, waste treatment R&D, mineral leaching and plant strain development, due to their perceived importance for the enhanced future competitiveness of Canada's natural resource industries. Although important initial steps have been taken to increase awareness of the opportunities presented by these areas, rather low government and industrial R&D expenditure levels in these areas belie the priority status accorded to them. In light of the long term and substantial commitment to R&D necessary before commercial objectives in these areas may be realised, experts suggest the need for significantly greater funding of underpinning research at universities to reinforce an under-funded research base, and more persuasive incentives for industry to invest in this research, to counteract a general long-standing resistance to investment in R&D.

In *Finland*, wood based industries, enzyme technology and analytic systems are considered to be the areas of greatest national opportunity for industrial exploitation of biotechnology R&D. No coherent programme with clearly identified priorities currently exists, as Finland pursues a more ad-hoc approach to biotechnology support in order to maintain flexibility; however, a limited number of strategic areas may be considered for targeted support under a national programme currently under preparation. In some critical areas where local technology is not sufficiently advanced, e.g. downstream processing, authorities anticipate importing technology and equipment, as well as supporting developments at home.

The *United States* and *Switzerland*, in accordance with long-standing political and economic principles, have deliberately not intervened in industrial biotechnology by establishing specific directions or objectives for R&D. The Swiss Government has, however, recently launched an initiative whereby special funds will be earmarked for concerted R&D programmes in areas of national importance. One proposal, based on consultation with universities and industry, and formulated by the Swiss National Science Foundation, related directly to biotechnology, and entailed plant biotechnology, enzyme technology and risk assessment research. The proposal, however, was ultimately rejected because of concerns that the research was too industrially-oriented and therefore an inappropriate topic for government-funded research.

Italy, which has been reviewing opportunities for industrial biotechnology, particularly within the sectors of agriculture and pharmaceuticals, has recently launched two research programmes under the auspices of the National Research Council (CNR) and the Office of the Ministry for Scientific and Technological Research (MRST). The "Finalised Project" on Innovative Biotechnology, administered by the CNR, will begin in 1988 and will entail support for precompetitive research in areas such as protein engineering, DNA probes, new vaccines, biosensors, cell cultures and fermentation processes. The anticipated cost is L 70 billion over five years. The Programme on Advanced Biotechnology established by MRST and proposed by the National Committee for Biotechnology will begin in 1988 and will entail support for applied research in areas such as monoclonal antibodies for diagnostics, DNA/RNA molecular probes, downstream processing, crop nitrogen metabolism and energy from biomass. The anticipated cost is L 400 billion over five years. Other priority areas for which grants will be made available include human genome sequencing and valorisation of agricultural by-products. *Belgium* and *Denmark* have refrained from identifying specific industrial priorities.

*
**

Amongst those governments which have launched industrial biotechnology programmes, a few have needed to address the rather fundamental problem of industrial inertia vis-à-vis R&D, complicating the task of choosing directions and selecting priority areas. Due to a variety of government initiatives and incentive plans designed to stimulate investments in biotechnology R&D, as well as general awareness-raising of the opportunities offered by biotechnology, however, some of these countries can point to improvements in their level of industrial R&D activity. For several other countries, the orientation of government strategies and the selection of priority areas may be directly attributed to the input of a dynamic and far-sighted industrial community already attuned to the opportunities and risks of biotechnology, and prepared to commit resources to future competitiveness in this area.

With respect to the particular industrial target areas selected as part of government biotechnology strategies, it is yet premature to judge the efficacy of the selection in terms of their enhancement of industrial strength and competitiveness; however, in many cases, governments appear to have selected targets judiciously, concentrating on competitive strengths and advantages, and in some cases, taking into account the expected future trends of biotechnology, e.g. a shift in pharmaceutical production away from therapeutics towards early diagnostics and vaccines, and expectations of major new opportunities in the agri-food area. Overall, although a few have remained rather insular, most of those governments reviewed have also demonstrated an active and growing attunement to developments and trends in biotechnology worldwide.

2. Biotechnology and agriculture

a) R&D developments and future prospects

It is still extremely difficult to estimate the precise impact that biotechnology will have on agriculture and related industries in terms of worldwide market size. Due to wide discrepancies in currently accepted definitions of what constitutes biotechnology, estimates for agricultural markets for the year 2000 range from $2-4 billion, to $21 billion, to $50-100 billion[11]. However, in the agricultural sector, biotechnology clearly represents a means for pivotal change. Among the more immediate impacts which biotechnology promises to exert will be improvements in the scope and efficiency of agricultural production. The application of new biotechnological techniques, such as rDNA, cell fusion and tissue cultures in the agricultural sector will make possible a wide range of improvements in the production of crops and livestock.

In the *plant sector*, for example, advanced cloning techniques based on tissue culture and plant genetic manipulation will contribute to, among other things, more efficient propagation, better selection of seeds, greater tolerance to environmental stress, resistance to pests and tolerance to pesticides. Although an insufficiency of knowledge about plant genetics and plant physiology is delaying the thrust of rDNA's expected impact on plant production, cell fusion and tissue cultures are already being used to breed more efficiently new and improved strains, and the application of rDNA techniques is expected to ensue.

In the *animal sector*, biotechnology will enhance livestock production through better diagnostic systems, improved therapy for and prevention of animal diseases by means of genetically modified products, such as hormones and vaccines, and finally, through the ultimate manipulation of the animal genome itself in order to achieve the expression of desired traits, such as greater nutritive content. Finally, in both the plant and animal sectors, biotechnology will provide a means for substantially increasing absolute levels of production.

The impacts of biotechnology will not, however, be strictly confined to improvements and modifications in traditional means of agricultural production; its direct and indirect effects will be felt in an *increasing convergence of agriculture and industrial practice*, creating a reorientation in relationships between and among agro-suppliers, farmers and the food processing industry, and introducing a new generation of science-based agricultural companies designed to exploit these techniques through products destined for a variety of agro-related markets[12]. These changes, based upon genetic engineering, tissue cultures and/or bioprocess technology, will create a number of significant public policy issues requiring review.

b) National policy issues

Biotechnology has the potential to boost agriculture production. A grave potential for conflict therefore exists between economic policies which create surpluses in certain agricultural commodities and the application of biotechnology in ways which might aggravate those surpluses. For example, the application of bovine growth hormone to cows increases milk production by up to 40 per cent. The large scale production of this hormone has been made possible by genetic engineering[13]. This points to the need to co-ordinate the directions and aims of biotechnology R&D with economic and agricultural policy, so that, with necessary adjustments on either or both sides, the products and by-products of biotechnology are channelled to markets where sufficient demand exists.

Another issue, concerning agricultural *trade and employment*, rests on the fact that biotechnology will render possible, through improvements in microbial and plant tissue culture, the industrial production of certain crops of vital importance to certain countries' farming communities. This capacity has already been demonstrated in the sugar market with the industrial production, through the conversion of starch, of high fructose syrups which compete with sugar. Although in the EEC, government intervention has succeeded in severely limiting production of these products through quotas and levies, this issue may become increasingly sensitive and pressing as biotechnology leads to other processes which challenge a number of traditional agricultural producers in other markets.

Furthermore, as biotechnology creates a shift in traditional *uses of land* in both developed and developing countries, allowing a wider range of crop production possibilities to farmers in climates where such possibilities were formerly limited by soil and weather conditions, certain changes in the international balance of trade will be inevitable. Developing countries may become more self-sufficient in the production of certain crops, while, on the other hand, developed countries may become able to compete with crops hitherto produced exclusively by the developing countries, e.g. cocoa. These changes, when possible, should be anticipated and reviewed, although at this time, particularly given the rapid rate of advances in biotechnology, predictions of their precise scope and nature would be premature.

Another major long-term shift in agriculture is the trend towards fewer and larger farms, with a consequent reduction of employment. As biotechnology will lead to productivity increases through factor saving effects (labour, land, energy, water), it is expected to accelerate this trend. In fact, it is the large rather than the small farms which will have the means to acquire and apply the new biotechnologies. The latter will need government help for this purpose[14].

Finally, as a result of advances in biotechnology, it will become increasingly possible to replace chemical pesticides with biological substitutes or ultimately to phase out pesticides altogether by genetically altering plants to resist pests and diseases independently[15]. However, along with the long-term prospect of an environment freer of chemicals, exists the concurrent possibility to markedly increase chemical applications, by utilising genetic engineering techniques to take hitherto herbicide-sensitive seeds and to modify them to resist and thrive under specific formulations of pesticides. Farmers would consequently be able to use more chemicals, more frequently on a larger variety of crops[16].

With this in mind, major multinational chemical and drug corporations, which have bought into more than 60 seed-producing companies during the past 15 years, are using genetic engineering in a race to create seeds that are herbicide-tolerant. Some company scientists contend that the development of such seeds will ultimately benefit the environment (in addition to increasing crop yields), by encouraging the competitive development of effective but less dangerous herbicides, and by providing research clues for the development of crops that thrive without chemicals. In the short run, however, a more probable scenario may be that "the commercial bounty reaped from biotechnology's herbicide-proof harvest may temper industry's pursuit of a nobler goal for the new science"[17]. Critics of biotechnology would tend to agree with this latter statement. Having either ignored or dismissed the potential role of biotechnology to improve environmental safety, these critics conclude that an inevitable impact of biotechnology on agriculture will be further *ecological damage*.

c) Government responses

Recognising the great potential offered by biotechnology to agriculture and the agro-industries, at least five of the 15 countries reviewed (Japan, France, Germany, the UK, and the USA) have launched fairly comprehensive or forward-looking programmes to support and accelerate R&D developments. In several of the other countries (e.g. the Netherlands, Canada, Denmark, Sweden, Finland, Belgium and Australia), interesting biotechnology research activities may be cited, although there appears to be an absence of the more sweeping or progressive strategies noted above. Although few governments have formally addressed the broad policy implications of biotechnology, in many cases, signs of a growing involvement of government with farms and agro-industry may already be witnessed, as competition in the agricultural sector relies increasingly on high-technology, and government support to small companies and farms through research, facilities, or finance is viewed as critical.

In *Japan*, the Ministry of Agriculture, Forestry and Fisheries (MAFF) is supporting a vigorous and comprehensive effort to develop and apply the techniques of biotechnology with stated goals of increased productivity in agriculture, forestry, fisheries and the food industry. Although the thrust of MAFF's biotechnology effort is devoted to integrated research for plant breeding, the collection and management of genetic resources and regional programmes, they have also instituted a scheme whereby industrial R&D in predesignated topical areas may receive government subsidies of 50 per cent. Government supported industrial research topics include bioreactor development, advanced breeding techniques through cell fusion and tissue cultures, development of new agricultural chemicals through biological means, diagnostic procedures for livestock diseases and utilisation of biological activators for the development of new fertilizers. MAFF co-ordinates this research to prevent overlaps, and any resulting property rights belong to the industrial participants. MAFF's industrial support scheme in biotechnology receives less than 10 per cent of the overall MAFF biotechnology budget, however, or approximately Y 258 million in 1986.

Additional research sponsored by MAFF includes the development of techniques for biomass conversion through enzymes and micro-organisms and analysis of the rhizosphere and development of its control techniques. MAFF has planned to establish an "Organisation for Research and Development of Specified Bio-Oriented Industry", financed jointly by the Government and private enterprise with the purpose of promoting the mechanisation of agriculture and biotechnology research for private sector enterprises related to agriculture, fisheries, and the food, tobacco and alcohol industries. The organisation would provide companies with interest-free loans and capital investment in biotechnology research, and would promote an R&D information service, international research co-operation and services for joint research with government bodies. The budget for fiscal year 1986 was estimated at Y 5.3 billion.

In *Germany*, the Federal Ministry of Agriculture (BML) has drafted a biotechnology programme as well, mainly for support of in-house research, based on an inquiry which it conducted regarding the current status of and future prospects for R&D, and the potential of biotechnology in agriculture and the food industry. Areas of research targeted as priorities include cell and tissue culture, genetic engineering, bioconversion and enzyme technology, and biotechnological control of physiological functions. A wide range of applications for these techniques has been identified generally as including the breeding, protection and growth of plants, the breeding, health and nutrition of animals, processing of agricultural and forest products and protection of resources, environment and safety. This programme has led to a co-ordinated initiative of BMFT and BML, entitled "Applied Biology-Biotechnology", under which specific priority research topics have been defined and supported. Research under this initiative is conducted in a thematically integrated way by public research institutes along with

29

industry. One research priority concerns production alternatives and new market outlets for traditional agricultural products. In addition to the issue of production alternatives, government officials in Germany have expressed an awareness and concern over other policy issues surrounding the impact of biotechnology on the agricultural community. For example, the need for better general forecasting of the social, environmental and economic impacts of biotechnology on agriculture, in order to predict its effect on farms and related industries, has been cited.

One issue of particular concern regards the role of government research and financial support vis-à-vis small seed-producing companies and small farms, whose dependence on government will become more acute as competitiveness begins to require rapid technological advancement, but where a traditional reluctance to invest in R&D still prevails. In Germany, agro-industries have traditionally relied on the public sector for R&D facilities; thus, the Government has always assumed some responsibility for their competitiveness. However, increasing demands for government assistance in agricultural R&D from various sectors may necessitate changes in agro-support policies. The BML has been developing and introducing new research methods, where those methods prove more efficient than traditional techniques. Advanced methods are already used in areas such as plant breeding and animal health, while in several other areas, modern techniques and equipment are lacking. In addition, the BML centres and industry might benefit from the expansion of links between them, which are currently rather limited.

In the *United Kingdom*, agricultural policy towards biotechnology has been less clearly defined than in either Japan or Germany; however, the UK is pursuing a deliberate general policy to sharpen the industrial relevance of government-funded research in agriculture. Hence, while more traditional and ostensibly less commercially relevant areas of research such as soil science for drainage problems are suffering cutbacks, funding for new areas of biotechnological research has increased. Priorities within biotechnology have been identified as new techniques of plant breeding, and enzyme engineering.

Experts have noted a shift to food research at the expense of agricultural research, and express concern that with the heavy emphasis on commercialisation and applied R&D in UK agriculture, more fundamental research with important long-term payoffs may be neglected, (e.g. research on genetic structure). Moreover, government agricultural research institutes are being encouraged to seek commercial sponsorship and research is being heavily directed towards perceived short-term markets, at the expense of more long-term fundamental research.

In *France*, the utilisation of biotechnology by the agri-food industries has been selected as one of two highest priorities of the Programme Mobilisateur. Deliberately avoiding the "classical" sectors targeted by biotechnology, namely therapeutic drugs, the French Government has chosen agriculture and the agri-food industries as a sector which has not been overly exploited, where French research is already competitive, and where innumerable opportunities for the development and utilisation of new products and processes exist.

The French agri-food research programme, entitled "Aliment 2000", has been designed to generate co-operative industry/university research in each of six theme areas. These are: identification of strains through monoclonal antibodies (one of the participant companies being interested in the utilisation of diagnostic methods for agri-food application); lactic acid bacteria research; mixed cultures research; enzyme use (particularly in complex situations e.g. involving low water content); on-line control of fermentation; and wine research, with the objective of better control of fermentation processes. These projects are all co-funded by the Ministry for Agriculture and the Programme Mobilisateur, with generally 50 per cent private financing. Generally, any resulting property rights belong to the participant company and no property rights or returns are demanded by the Programme Mobilisateur.

French authorities assert that European collaboration could be pivotal to the development of agricultural and agri-food research, and are strongly in favour of promoting such co-operation. Seed technology was referred to as one potentially attractive area for collaboration, principally through EUREKA, while co-operation in more fundamental aspects of agricultural biotechnology, under the aegis of the *Commission of the European Communities*' Biomolecular Engineering Programme (BEP), and subsequent Biotechnology Action Programme (BAP), has already received considerable support.

In the *Netherlands*, where conventional plant breeding has traditionally been excellent and government institutes bear the responsibility for developing new plant varieties, there has been some reluctance to accept and adopt non-conventional methods, and agricultural institutes have been criticised by experts as highly inflexible. However, seed companies are gradually beginning to adopt new techniques on their own. Under the Dutch biotechnology programme, the Ministry of Agriculture and the Ministry of the Environment are jointly supporting R&D on farm waste degradation, e.g. the bioconversion of manure into fertilizer marketable in the Third World.

In the *United States*, initiatives to promote biotechnology research in the plant sciences have recently broadened, in response to the growing importance of improving agricultural research in order to maintain a competitive position in export markets. Significant efforts have been made to encourage an increase in federal support for the long-term research in plant genetics and physiology which will be necessary before the opportunities presented by biotechnology may be fully exploited and plants with agronomically useful traits may be engineered. For example, the National Science Foundation hopes to integrate biotechnology into its established biological research in the plant sciences, whereby, at the cost of $7.5 million annually, it would seek to train hundreds of post-doctoral researchers in multidisciplinary plant research. Furthermore, in proposals to the White House Office of Science and Technology Policy, the National Science Foundation, the Department of Agriculture and the Department of Energy have pinpointed several biotechnology-related research areas for further support, focusing on: microbial ecology and genetics in the rhizosphere, chemistry of complex plant carbohydrates, molecular biology and genetics of plant cells, modelling of agro-ecosystems, and information technology. The Department of Agriculture is taking independent steps as well to expand its work on plant biology, opening a Plant Gene Expression Center near San Francisco, with expected annual support of approximately $6 million[18].

In *Canada*, which boasts a strong tradition in agricultural research, there exist several excellent biotechnological research efforts which could be further developed through a greater concentration of funds and manpower. These efforts include crop improvement using embryo culture and sexual and somatic hybridisation, and gene transfer between different plant species. Specific goals for this research include a frost-tolerant alfalfa, salt-tolerant flax, tobacco and sugar beet, and genetically improved oilseeds, cereals and horticultural crops. However, an insufficiency of research on each topic has been pointed to as a serious possible hitch in developments. For example, only three research groups are focusing on salt tolerance in plants and only two groups are working on frost tolerance. This reflects a general situation in plant biotechnology research in Canada, characterised by high quality but small and isolated research groups. Although more than 100 research groups are working in this field, their efforts have been deemed "scattered" and full of "gaping holes"[19].

Another weakness in Canadian agricultural biotechnology is a lack of significant research sponsored on the application of biotechnology to food processing. The recent establishment of a food research centre by the Department of Agriculture will help to address this problem, with work on food processing techniques, enzyme research, bioreactors, etc. This centre has been built with the intention of attracting the food industry to its expertise and facilities, and hence stimulating strong interaction. In addition, some provincial initiatives have been taken in this field, such as the establishment by the Government of Quebec of a venture capital company called BioAgral, to support biotechnology developments in the food industry, and the operation of the Canadian Food Products Development Centre by the Province of Manitoba. In the area of livestock production and veterinary biotechnology, considerable research is being done in Canada with the objective of reducing production costs through increased yields, disease protection and genetic selection.

*
**

Despite a clear recognition that biotechnology will impart new benefits and opportunities to agricultural production and related activities, and hence the programmes and initiatives noted above, surprisingly little attention or comprehensive thinking has been directed at government levels towards reconciling these anticipated benefits with the broader policy issues concerning agricultural production. The EEC's Concertation Unit for Biotechnology in Europe (CUBE), however, has made some strides in this direction. Speaking in terms of a "New Partnership between Agriculture and Industry", CUBE has studied ways of reorienting agriculture through biotechnology to mitigate surpluses and enhance agricultural employment by finding alternative uses for surpluses or land, e.g. designing crops for market needs, developing agriculture as a supplier of high value raw material to the manufacturing industry, etc.[20]. Although some of the options considered by CUBE are not presently viable economically, and none even approach an absolute solution to the surplus problem, the EEC will conduct studies and support research to pursue the more promising opportunities. A sizable portion of the EEC's ECU 15 million Biomolecular Engineering Programme budget (1982-86) and the subsequent Research Action Programme's ECU 55 million budget (1986-89) supports research relevant to the agri-food sector. Both programmes have received relatively strong support from the EEC's Member States.

3. Biotechnology and public health

a) R&D developments and future prospects

For nearly a decade, experts have referred increasingly to the "revolutionary" effects which biotechnology would impart to health care, emphasizing the vast potential conferred by recombinant DNA and hybridoma technology for improved capabilities in the production of therapeutics, in diagnosis and in prevention. By providing a greater understanding of physiological processes and the functioning of the immune system, these two discoveries have been hailed for the promise they extend to produce highly targeted therapeutic drugs in adequate supply, and to diagnose, prevent and ultimately cure a wide range of hitherto intractable diseases.

Although in some cases, early hopes and expectations based on these techniques have met with temporary disappointments or deferrals, e.g. anticipations that recombinant DNA produced interferon might provide a cure for cancer, tangible results of rDNA and hybridoma technology have indeed emerged in the health care industries. In addition to rDNA produced human insulin, one of the earliest rDNA products ever to reach the market, monoclonal antibodies have demonstrated commercial viability as diagnostic kits whose use, *in vitro*, allows them to circumvent the lengthy regulatory procedures which have traditionally confronted pharmaceutical products. Other products still in the pipeline but which, according to all available evidence, are approaching the stage of commercialisation, include a host of vaccines, blood factors, immunoregulatory and modulating agents, and, in the more distant future, neuropeptides.

The anticipation of major scientific and commercial developments in pharmaceutical production due to biotechnology is evident within most pharmaceutical firms, where the development of health care products, based on biotechnology, has been the object of intense competition for a number of years. Several governments are directly or indirectly supporting these developments through various industrial policies mentioned earlier, as well as through their support for underpinning biomedical research; however, in some countries, particularly those which have lagged behind in applications of biotechnology to the development of major pharmaceuticals, policy-makers have alluded to a deliberate effort to support R&D in "new" as opposed to "classical" areas of pharmaceutical production. In the case of France, for example, policy-makers have emphasized the development of new diagnostics and vaccines as opposed to therapeutics. This may be either a direct result of what has been deemed a general scientific trend in biotechnology away from therapeutics to prevention and diagnostics, or the outcome of general policy to avoid those heavily pursued areas in which international competition may already be too fierce.

In addition to the manifold opportunities which it presents to pharmaceutical product development, biotechnology will impart to health care the truly revolutionary possibility of performing human gene therapy, to alleviate genetic defects caused by hereditary and non-hereditary diseases. The first efforts in this field will involve the insertion of genes into the bone marrow cells of patients, *in vitro*, and their reinsertion into the patient's body, where the non-defective genes should theoretically multiply and achieve expression. Although the first clinical trials for human gene therapy in the United States were expected to begin in 1986, substantial advances are still needed in the development of retroviral vectors before such trials may be successful. Diseases susceptible to this form of treatment would include hereditary anaemias and immune deficiencies, as well as haemophilia.

b) National policy issues

Although the demonstrated and anticipated advances in health care engendered by biotechnology will clearly contribute to the public good, many of these developments will present some challenging public policy issues which deserve consideration. Some of these issues have no precedent, while others have been raised before, but re-emerge with greater urgency as biotechnology rises to the fore. Amongst those unprecedented issues are the questions arising from the anticipated shift in medicine from treatment to diagnosis and prevention. Increasing opportunities for early diagnosis of viral diseases, e.g. AIDS or genetic defects will alert people to their status as carriers of certain diseases or to their tendency to develop a certain disorder long before symptoms may become evident, and in many cases, well before comparable or proportionate improvements in therapeutic treatments may exist. In cases of infectious or hereditary disease, such information might contribute vitally to its containment; in other cases, however, the tests will simply provide warning of a tendency to develop a certain disease or disorder without there necessarily being a treatment or cure. The increasing capabilities for early diagnosis raise certain ethical questions, such as who should be tested, and what should they be told about the results of the tests?

It is also conceivable that, with the envisaged shift in medicine from treatment to early diagnosis and more significantly to prevention, the economics of public health could be significantly modified as a result of biotechnology. Given the magnitude of health expenditures in OECD countries (approximately 7-11 per cent of GNP) and their pressures on national budgets and budget deficits, this question certainly deserves examination.

The growing prominence of biotechnology imparts greater urgency to several perennial issues facing public health administrations. One of these issues to which biotechnology adds a new dimension is that of "orphan drugs", pharmaceuticals for which urgent demand exists, but where anticipated markets are so small as to act as major disincentives to private investments in R&D. The "orphan drug" issue becomes more pressing as biotechnology may provide the technical potential to develop new drugs for the treatment or prevention of rare genetic disorders and rampant infectious diseases in the Third World, but which for reasons of profit and loss may be subject to neglect.

Another issue which will conceivably arise in the context of biotechnology will be: who should decide and what should be the criteria for selection of patients for expensive treatment for rare genetic diseases? This issue has some precedent in recent history of clinical innovation, from renal dialysis in the 1960s to the first heart transplants, and more recently in the use of artificial hearts.

c) Government responses

Government policies in the OECD, with regard to biotechnology and health care, generally exist in the form of financial support of underlying biomedical or life science research, and support of innovation and R&D by the pharmaceutical industry, through various fiscal incentives, loans, etc. Several major efforts have been made to support "orphan drug" development: a federal Orphan Drug Act was passed in the United States in 1983, whereby eligible drugs (those with patient populations of less than 200 000) may be entitled to federal support in the form of tax credits, exclusive marketing rights, etc.; in Japan, the Bureau of High Technology Promotion is supporting orphan drug development by Japanese companies, concentrating on pharmaceuticals for muscular dystrophy and enzyme deficiencies, through R&D tax breaks and subsidies. Australia, similarly, has fostered the development of a malaria vaccine through the Australia Industry Development Corporation's "national interest" provision, investing A$9 million of federal government funds in a development project. In *Germany* as well, the Federal Ministry for Research and Technology supports the development of a malaria vaccine. However, these efforts are not common in most of those countries reviewed, and even where they exist, e.g. in the United States, some pharmaceutical firms argue that the incentives for long-term R&D on Orphan Drugs are insufficient, and as a result, some firms have withdrawn from R&D on such products[21].

However, generally speaking, very few public health and medical research administrations appear to be reviewing the potential consequences of biotechnology R&D, studying the ways in which they might influence its direction, or considering policy changes to prepare their large and often inflexible public health systems for the consequences of this new technology. Furthermore, in a majority of countries surveyed, no concerted effort appears to exist to facilitate transfer of research developments in biotechnology from government laboratories to hospitals. In view of the fact that already a decade has elapsed since the impacts of biotechnology were predicted and deemed "revolutionary", it seems rather surprising that some of these issues have received so little attention in policy circles, perhaps even less than that given to the broad policy implications of biotechnology in agriculture.

However, although these findings appear to be true for the majority of countries surveyed, some notable exceptions do exist. In the *United States*, for example, the National Science Foundation (NSF) and the National Institutes of Health (NIH) assume a fairly broad and active role in reviewing developments in biological sciences, and considering policy issues. Beyond its massive support for fundamental biomedical research and training, the

NIH was the first health organisation to establish "Points to Consider in the Design and Submission of Human Somatic-Cell Gene Therapy Protocols", and discussions take place at high levels regarding the various aspects of the agency's influence on the direction and evolution of the field. The Food and Drug Administration has played an active and important role in supporting the development of health-related biotechnology products, accelerating formerly lengthy approval procedures for biotechnology-based products, and the Office of Technology Assessment has published a considerable number of policy studies related to biotechnology and public health[22].

Japan's Ministry of Health and Welfare, where most biotechnology research falls under a national cancer control programme, is also pursuing guidelines for human gene therapy, and has organised a Council for Human Health to examine the NIH's "Points to Consider...." In the *Netherlands*, a Health Committee for Biotechnology has recently been established, and the Dutch Cancer Institute has instituted a training programme designed to acquaint hospital workers with new biotechnologies, e.g. rDNA work, oncogene testing, etc.

4. Energy and the environment

The application of biotechnology to the exploitation and the enhancement of available *energy* sources may occur through several routes, including the conversion of biomass through fermentation processes to biofuels, e.g. methane and ethanol, microbiological removal of sulphur and sulphides from coal, and enhanced oil recovery.

Of these options, the conversion of biomass has received the most attention as a possible alternative, or more likely, a supplement to the utilisation of non-renewable fossil fuels, and research on this topic has received government support in several OECD countries. According to conventional wisdom, however, biomass does not represent a viable alternative, at least for the present time, because of process inefficiencies and uncompetitive costs. Not only does ethanol production require enormous quantities of raw material and entail tremendous costs for collection, transportation and storage of these materials, which exceed the value of the energy output, but for most starch crops, negative energy balances exist in the production process. Even if advances in genetic engineering and enzyme immobilisation should lower the costs of fermentation in the near future, fossil fuels at present prices (March 1987) would remain a more cost-efficient energy source.

Despite the short-term disadvantages of biomass as an alternative energy source for most OECD countries, several governments have provided support for bio-energy R&D, with a view to reducing dependency on fossil fuels for domestic energy consumption, or to developing biofuel conversion systems for export to developing countries. The most concentrated efforts to develop biomass conversion systems for domestic use in the OECD have occurred in *Japan*, where the incentives to reduce oil dependency are particularly keen. Both the Ministry for International Trade and Industry and the Ministry for Agriculture, Forestry and Fisheries are providing vigorous support for R&D in this area.

The *United States* Department of Energy also supports a wide range of activities related to biotechnology, primarily in the conversion of biomass to fuels or energy intensive chemicals which can replace petroleum-based feedstocks. Areas of particular interest include lignocellulose processing for ethanol production, advanced bioreactor design, biocatalysis technology and biomass productivity.

Advanced research activities in biomass conversion technology may be found in the *Netherlands* as well, where the anaerobic digestion of waste has been applied to farm manure and to liquid waste from sugar refineries, and the generated methane subsequently used for fuel. In *Switzerland*, on a smaller scale, the Swiss National Science Foundation and the Swiss

National Foundation for Energy Research have supported bio-energy research through which a pilot programme to produce biogas from fermented manure was successfully completed, and the gas subsequently used to power farm equipment. Both Switzerland and the Netherlands have sought export opportunities in developing countries for biomass conversion systems. Other countries supporting biomass conversion technology include *Denmark*, where a comprehensive research programme in bio-energy and especially on biogas has been developed in recent years, and both *Finland* and *Germany*, where research efforts are active in the conversion of lignocellulose and in the use of anaerobic techniques for waste treatment and biogas production. Canada has also expressed an interest in bio-energy and has developed numerous pilot facilites.

In *France*, the *United Kingdom*, *Australia* and the *EEC*, only minimal interest in bio-energy has been expressed, while somewhat greater interest exists in *Italy*.

*
**

Research and development in biotechnology has a vital role to play in the domain of *environmental* protection, namely through the use of biosensors as pollution monitoring devices, the development of improved waste treatment systems, and the elucidation of scientific phenomena in order to improve risk assessment and risk management capabilities, and hence enable the safe environmental release of genetically engineered micro-organisms.

Biotechnology has traditional applications in waste treatment such as aerobic process treatment of solids and waste water, anaerobic processes used in sewage treatment for sludge stabilisation, and composting. Possible future applications may include reclamation of useful substances from waste by means of genetically engineered micro-organisms, and use of such organisms to effectively decompose crude oil or highly toxic substances. Experts are also considering integrated waste treatment processes, involving microbiological pre-treatment of wastes, followed by chemical or physical decomposition processes.

Several OECD countries have been pursuing biotechnology research and development in improved waste treatment processes, notably the *Netherlands*, *France*, *Japan* and *Germany*. However, despite significant developments in waste treatment processes throughout the OECD, their adoption by industry has been described as problematic in many countries reviewed, because of lax regulations which encourage the payment of fines by industry for waste emission rather than the utilisation of efficient industrial treatment systems.

In the *Netherlands*, where highly concentrated industrial activity and high population density have necessitated strict regulation of waste disposal, industrial incentives to develop and utilise efficient waste treatment systems have been effectively introduced through legislative means. Companies such as Gist-Brocades utilise and are striving to market advanced anaerobic waste water clean-up processes. As part of its efforts to promote and support improved waste treatment technology, the Dutch Government supports research in soil biodegradation, and the Ministry for the Environment has been involved in joint projects with industry and agriculture, in the latter case, to develop systems to convert farm waste in small fermentors into marketable fertilizer for developing countries.

In the *United Kingdom*, on the contrary, notwithstanding the R&D efforts of several small companies and regional water authorities, the utilisation of new processes by industry occurs rarely, due to a less stringent regulatory climate and weak incentives for efficient industrial clean-up. In *Canada* as well, although waste treatment R&D exists, the regulatory climate has been relatively lenient in this area, and little industrial diffusion of new process technologies has occurred.

In the *United States*, many experts have described the development and industrial utilisation of new waste treatment technologies as inadequate, and emphasize particularly the several critical impediments to the development and diffusion of new technologies involving genetically engineered micro-organisms targeted for specific forms of pollution, capable of survival in highly toxic environments, or possessing other novel capabilities. These experts contend that the development of biotechnology products in waste treatment are lagging behind product development in other sectors of industry, and attribute this fact to field testing and risk assessment hurdles, liability issues, concern over possibly negative public reactions, and lack of industrial incentives to develop efficient new waste treatment processes. Industry dœs not perceive a need or a market for new pollution control products, as the payment of fines for environmental pollution still remains more economical than industrial clean-up[23].

Government support of industrial R&D in waste water treatment has been considerable in both *Japan*, where in addition to MITI's activities, the Ministry of Construction has launched a five year, Y 5 billion project on waste water treatment through biotechnological processes, and in *France*. In *Switzerland*, efforts in waste water treatment have been supported at the Federal Institute for Water Resources and Water Pollution Control in Zurich. Other Swiss R&D efforts in waste treatment include work on optimisation of aerobic digestion processes for secondary sludge treatment, and microbial degradation of nitrolotria-cetate, although these activities have been rather minor in scale.

In addition to regulatory disincentives in many OECD countries to the development and utilisation of innovative new waste treatment systems by industry, a significant disincentive to innovation revolves around the existing gaps in scientific knowledge which prevent accurate assesments of potential risks. Without acceptable criteria and methodologies to evaluate the risks posed by the environmental release of genetically engineered micro-organisms and consequent regulatory uncertainties, many companies are unwilling to embark on new avenues of research. This presents a major hurdle to developments in biotechnology in all areas involving environmental release of genetically engineered micro-organisms, waste treatment R&D included. The imperative for advances in risk assessment before the opportunities presented by biotechnology may be fully and safely exploited cannot be overstated.

Given the vital importance of biotechnology as an instrument to help monitor and clean-up the environment, the scarcity of efforts at the government level to support and encourage research activities in this area is rather disturbing. Environmental ministries in most OECD countries more often than not seem to assume a defensive posture vis-à-vis biotechnology, concerning themselves with how to regulate against the hypothetical environmental risks posed by biotechnology, rather than how to exploit the many ways in which biotechnology might benefit the environment. Moreover, because the fruits of such research, when genetically engineered micro-organisms are involved, may not be reaped until advances in risk evaluation capabilities are made, the scarcity of research efforts in risk assessment is a point of equal, if not greater, concern.

Exceptions exist, notably in the *United States*, where the Environmental Protection Agency supports a new biotechnology research program, under which several research activities related to risk assessment are being conducted. Research activities include projects examining microbial survival and growth, transfer *in situ* of genetic information, dispersal, biological containment and the environmental and health effects of genetically engineered micro-organisms. In *France* as well, the Ministry of Agriculture, along with the Ministry for the Environment, is beginning to support risk assessment research. The French Government is also supporting the application of biotechnology to pollution control, e.g. the use of biosensors for on-line control of organophosphate pesticides. The *German* Government, through the Federal Ministry for Research and Technology, plans to introduce a programme supporting risk assessment research, and research supported under the *Commission of the European Communities'* biotechnology programme includes some work on risk assessment R&D as well. However, notwithstanding some significant examples of research activities in the field, the overall extent of risk assessment research in the OECD appears inadequate given the urgent need for substantial and rapid developments. This may be largely due to the difficulties experienced by governments and research funding agencies in attracting talent to the field.

III. COMMERCIAL EXPLOITATION OF UNIVERSITY AND GOVERNMENT RESEARCH

1. University/industry research alliances and implications for basic research

The close proximity of much fundamental biological research to industrial application, both in product and process development, has spawned a proliferation of industry/university research agreements which provide industry with a window on and rights to exploit research, and provide universities with additional sources of funding. This proliferation of industrial/ university research agreements has incited a debate regarding the possible implications for basic research, and the appropriate role of governments in this context. Among the issues facing governments are the following: will the increasing orientation of university researchers towards co-operation with industry impinge upon basic research and undermine freedom of inquiry and dissemination of knowledge? If not (though evidence suggests that already, in many instances, communications breakdowns have occurred amongst biotechnologists in academic circles), is the level and type of industrial support sufficient to justify further reductions in government support of biotechnology research in universities?

According to evidence found in the United States and most likely applicable to the OECD countries in general, university/industry agreements can be a "positive sum game, one in which both sides benefit", provided contracts are negotiated and drawn carefully, to protect academic freedom, patent rights and timely publication. However, according to these same findings, company support of university research cannot supplant government support, for although industry supports a larger proportion of university research in biotechnology than it supports in the average scientific field, most of its support tends to be shorter in duration than government support (2 years or less), and thus more applied[24].

Governments must continue to support fundamental research in biotechnology, particularly in areas such as agriculture and many aspects of health, where commercial applications depend on much greater understanding of fundamental processes, and where industry might be less inclined to provide adequate support. Basic research in universities has been the foundation on which most commercial developments have been based in biotechnology and has led to an unusually large number of patent applications relative to those generated from research performed in company laboratories. Therefore, there is a strong argument to be made that governments ensure the strength of basic research capabilities in universities in order to safeguard the underpinnings of future commercial developments. However, governments might also consider providing incentives for companies to support longer-term research with significant industrial potential, such as the initiative launched in the United Kingdom to encourage industrial consortia to conduct research in plant molecular genetics.

2. Policies to facilitate university/industry co-operation

As university/industry links have been proven a pivotal factor in the development of biotechnology, governments have begun to adopt policies to foster such alliances, allowing companies to exploit leading edge research and providing universities with a new source of needed financing and exposure to practical industrial problems. In order for such research alliances to be successful, however, the following conditions should exist:

— *Excellence of Research* (world class teams whose research expertise corresponds with the objectives and profiles of domestic industry);

— *Communication and Information Exchange* (enabling companies to formulate goals and recognise problems in terms of opportunities offered by biotechnology and by the research expertise available to them);

— *Flexibility* (to pursue different approaches to and modes of co-operation, including consultancy services by academic personnel to industry, and negotiable property rights such that agreements may be reached to the satisfaction of both academic and industrial parties, etc.);

— *Industrial R&D Investment Incentives* (Where market pull is lacking, a variety of "government push" strategies have been resorted to, such as government subsidised precompetitive R&D programmes, low interest loans, etc. Government procurement has also been suggested as a possible means of stimulating investment. Often, these incentives are conditional on requirements to conduct co-operative research with universities or government laboratories[25].

In countries where some of these conditions have not traditionally existed, governments have introduced policy measures designed to achieve them. In certain cases, however, it may be argued that a more efficient solution would be for companies to simply invest in or license foreign research. In a few cases, governments have attempted to ensure that commercial liaisons exist with public research, particularly when established companies demonstrate indifference towards such research and venture capital start-ups have not filled the gap, by establishing new firms to prevent commercial neglect of potentially important research discoveries emerging from public laboratories.

Amongst those countries reviewed, links overall appear to be quite prevalent in the United States, Sweden, Germany, and the Netherlands. In the United States, a prevailing climate conducive to commercial exploitation of university research existed prior to the emergence of biotechnology, and links were fostered quite readily. In Sweden, Germany and the Netherlands, fairly vigorous government policies were employed to improve existing conditions for commercial exploitation of public research in biotechnology. In Austria, Finland, France and the United Kingdom, efforts to expand university/industry contacts have had significant success, while in Belgium, Denmark, Canada and Italy, links are rare, and in Japan and Switzerland, generally limited in scope.

In the *United States*, exploitation of university research underlying biotechnology resulted not from any direct or deliberate government policy, but was rather facilitated serendipitously by a healthy investment climate and a dynamic venture capital market which, along with a good measure of entrepreneurial spirit, gave rise to the celebrated "new biotechnology firms" established to capitalise on leading edge university research. These companies improved and exploited the new techniques of biotechnology, using them as tools for product development and producing the first generation of new biotechnology products. They also provided a smooth transition between university research and industry, as mature firms began to invest in their research, license their products, or acquire their research facilities and staffs. This transition, as well as direct links established between large firms and universities, has been abetted by a fiscal policy which favours R&D investment[26], open communication and flexibility, and an outstanding research capability in the biological sciences. The exceptional extent of scientific excellence in the United States has been fostered by years of highly concentrated funding of biological and medical research, the magnitude of which surpasses that of any other country in the OECD.

The *Swedish* Government has recently introduced several policies to encourage the development of links between the academic and industrial communities in biotechnology, as much to encourage the flow of industrial funding into an underfunded academic research community, as to encourage industrial exploitation of university research. Links between government, industry and universities have traditionally been good, but the increasing attraction of industrial research to the academic research community and young graduates may jeopardise the balance between them. In order to make academic research more attractive financially, the Government has introduced policies to facilitate and to promote industrial employment opportunities for academics. For example, new legislation was introduced which recognises remuneration of academic research personnel for industrial research work. This legislation was mainly intended to increase co-operation between academia and industry. Many experts, however, contend that this legislation was in direct response to the drain of scientists from academia to industry. Currently, most of Sweden's finest academic researchers in biotechnology perform consultancy work for industry on a regular basis.

Similar to the experience of the United States, Sweden has recently witnessed a proliferation of new biotechnology firms, mostly university spin-offs, regarded positively as another means of encouraging professors to remain affiliated with universities. However, unlike the US venture capital start-ups, many of the Swedish start-ups, while strong scientifically, lack the business and management expertise necessary for survival. A major concern expressed by both industrialists and government officials regarding these small companies is that, rather than collaborating with established industries, they might try to set themselves up as competitors, and refuse access to their research. In a country where severely limited manpower is already a serious constraint to development of biotechnology, a further dispersion of this critical resource may inevitably entail serious repercussions.

Encouraging the flow of ideas from industry to universities as well, Sweden has introduced an "Adjunct Professorship" scheme, first conceived of by pharmaceutical companies in the field of biotechnology, whereby industrialists undertake part-time teaching in universities. In addition to exchanges of research personnel between academia and industry, a large number of industrial grants and contracts supporting university research exist, notably at the Universities of Uppsala, Lund, and Umea. The University of Umea is also performing contract work with Kabigen on cloning and expression of Factor VIII, a blood clotting agent. Finally, the National Swedish Board for Technical Development (STU) has a policy to ensure the relevance of academic research to industry, and has thus been funding industrially-relevant basic and generic applied research at universities, and working closely with committees of academic and industrial membership to define strategic areas for support.

The climate in *Germany* for industrial exploitation of university and institute research is excellent, but unlike in the United States, Sweden and the United Kingdom, few biotechnology start-up companies exist in Germany, and thus the emphasis of governmental support to stimulate the exploitation of biotechnological research appears to be on established industry. Currently, nearly every major German chemical company has established collaborative biotechnology research programmes with German universities. In addition, to aid the smaller and medium sized firms "tap the latest developments in the field", the chemical industry association (Verband der Chemischen Industrie) will organise an arrangement to promote enhanced communication between them and the Society for Biotechnology Research in Braunschweig (GBF)[27]. Although links between Germany's chemical and pharmaceutical firms and university research laboratories have traditionally been strong, considerable efforts were made to reinforce and expand them, particularly in the area of biotechnology, after a $50 million research agreement was concluded by the Hœchst Corporation with the Massachusetts General Hospital in 1981.

Building on the coherence established between areas of academic research expertise and industrial research interests, the Federal Ministry for Research and Technology (BMFT) has adopted schemes to catalyse an increased number of working relationships between university and industrial researchers. BMFT actively encourages and supports projects which join firms and universities or institutes, provided these projects demonstrate a critical mass and coherent theme, which corresponds to the Ministry's biotechnology priority areas. Although single group projects in priority areas are still eligible for support, preference is given to projects based on cross-sectoral links. In such arrangements, BMFT subsidises the research group 50 per cent of the total project cost in order to defray the share of the costs incurred by the university, deliberately imposing the balance of the costs on the company to ensure industrial efficiency. The stated rationale of this scheme is to provide university researchers with an incentive to pursue industrially applicable research projects, by rendering them newly eligible for support through BMFT, more richly endowed with available funds than the German Research Society (DFG). Nevertheless, many such research scientists still prefer to maintain independence from industrial research.

A great deal of co-operative research takes place at the GBF, Gesellschaft für Biotechnologische Forschung, where 90 per cent of operational costs are financed by BMFT and 10 per cent by the State of Lower Saxony, and research projects are financed by industry. Also, the chemical industry has projected a plan to provide money to GBF in order to finance seminars and projects in conjunction with universities.

Perhaps the most notable initiative taken by BMFT to foster links between industry and public research in Germany, has been the establishment of "Gene Centres", built around university and Max-Planck facilities. Financing of these centres is provided 50 per cent by BMFT, with private companies covering the rest, except when universities are involved, in which case the Länder (state governments) also contribute to the funding. Gene Centres exist at Cologne, where Cologne University is collaborating with the Max-Planck Institute for Plant Physiology and Bayer AG, at Munich, where the University of Munich is collaborating with Hœchst AG and the Max-Planck Institute for Biochemistry, and at Heidelberg, where BASF AG co-operates with the University of Heidelberg.

In *The Netherlands*, government policy towards biotechnology has deliberately sought to foster greater links between Dutch research and Dutch industry, most notably through establishment of a new scheme, targeted towards biotechnology, called the "Integrated Applied Projects" programme (ITP). Under this scheme, which fosters research programmes aimed at producing industrial applications, the Ministry of Economic Affairs subsidises research costs incurred by industry 45 per cent, provided a university or research institute participates in the research. This scheme has been budgeted 50 per cent of the entire Innovation Programme (IOPb) budget, amounting to Gld 35 million.

Before this programme was instituted, Dutch companies were conducting much biotechnology R&D in the US, via start-up companies such as Cetus and Genentech. More industry/university co-operation is now taking place with Dutch research groups, a development which has been attributed to the success of the ITP Programme[28].

Changing attitudes are also an important factor in increased collaboration, particularly in the non-technical university community, where, although never legally forbidden, contact with industry has traditionally been discouraged. However, a combination of serious budgetary cutbacks and consequent need for new sources of finance, along with a dynamic industrial interest in getting a window on university research, has paved the way for a variety of new co-operative arrangements between industry and universities. Currently, most of Holland's renowned scientists perform consultancy work for industry.

Despite a definite proclivity towards support of applied, or generic applied research in biotechnology, basic research is theoretically protected through the following mechanisms: researchers are expected to divide their time 50/50 between basic and applied research; "basic" work in biotechnology is less vulnerable because of the long-term applicability to industry; ZWO, the basic science research funding agency, although inclining somewhat towards applied science, has made biotechnology a priority area for support, and the biotechnology advisory committee is mandated to help support basic research. In addition, 15 per cent or Gld 10 million of the Innovation Programme budget has been designated to support basic research.

In *France*, despite significant progress, two deep-rooted problems prevent more successful co-operation between industry and public research. One is the persistently adverse attitude of academic researchers towards co-operation with industry. The other, a unique problem directly due to the barriers in the French educational system between engineering and research, is the lack of familiarity amongst directors of research in French industry with research. Trained as engineers with little or no research background, these industrialists have difficulty in presenting viable research proposals to academics or in communicating effectively with them.

In the *United Kingdom*, links between industry and academia have been traditionally weak, and outstanding research achievements have often been exploited abroad, a notable example being the foreign exploitation of Köhler and Milstein's discovery of monoclonal antibodies at the Medical Research Council's molecular biology laboratory at Cambridge. Experts attribute this problem largely to industrial averseness (with the exception of a few large firms such as ICI) to investment in high risk research. Both academics and government officials refer to the greater readiness of US and Japanese companies to develop liaisons with UK researchers and the vigour with which the latter recruit them. Furthermore, support of research activities at universities and research institutes by British firms has been deemed inadequate.

In order to improve the climate for co-operation and to stimulate British industry into support of long term research, particularly that conducted at universities and research centres, the Department of Industry's Biotechnology Unit (BTU) and the Science and Engineering Research Council (SERC) have been working together to build "an invisible network of contacts" and to identify specific project areas of industrial relevance for priority support. By sharpening the industrial relevance of government funded research, making industry more aware of the opportunities provided by biotechnology, and finally taking the actual initiative to establish tangible research programmes, both BTU and SERC hope to improve transfer of research results to British industry, and at the same time improve the health of the academic sector by introducing new sources of funding.

SERC's Biotechnology Directorate, run by a management committee half of which consists of industrial membership, has instituted schemes to encourage industrial support of commercially applicable university research, e.g. through "clubs" (a scheme unique to SERC's biotechnology and electronics directorates) and co-operative grants. According to these programmes, a company must contribute 50 per cent to the costs of research in order to obtain resulting property rights; however, a contribution of less than 50 per cent entitles the firm to negotiate for preferential royalty fees, advance access to results, etc.

BTU has introduced a scheme to encourage industry to sponsor fundamental research in universities and institutes, whereby industrial funds for research projects are matched by the Government, provided three or more firms contribute, and provided industry contributes 50 per cent of the research costs. The rationale for this provision is that a commitment of 50 per cent by industry ensures serious industrial commitment to and oversight of the research. One result of this scheme has been a programme at the Institute for Biotechnological Studies (IBS), where BTU, along with seven firms, is jointly funding a five year programme on second generation bioreactors.

Since the programme at IBS was instituted, BTU has enthusiastically endorsed the concept of building industrial consortia for collaboration on pre-competitive R&D. Research consortia are viewed as an appropriate and expedient means of furthering technological developments, given conditions such as the requirement of a major long-term commitment of resources which individual companies might not be able to provide, research which is risky but which could generate enabling technology of general applicability if successful, and a dispersal of skilled manpower or necessary equipment. Moreover, collaborative research consortia should lead to the establishment of a broader skill base, which would be necessary even if the needed technology were to be obtained by licensing, and should foster greater overall awareness of opportunities among participants. However, pre-competitive R&D consortia entail the pitfall that companies might only be willing to collaborate in areas unlikely to generate commercial results, in which case government expenditures would be wasted.

The UK has focused on agricultural and food research, which seems particularly well suited to collaboration, as much of it is still at a precompetitive stage, but where the potential for exploitation of the UK excellent agricultural and food research base promises to revolutionise agricultural production and the food industry. Thus far, one project involving industry, the Durham and Warwick Universities and two Agricultural and Food Research Council (AFRC) institutes has been set up to investigate plant molecular genetics. This initiative sprang from a joint effort by DTI, SERC and AFRC to identify specific project areas in plant biotechnology for support. The difficulty of drawing together funding from three Government agencies reduced the scale of the initiative and DTI launched the Agri-Food programme alone in 1986. The degree of activity which the Agri-Food Scheme will generate amongst companies remains to be seen.

In addition to its efforts to transfer enabling research from university and research institute laboratories to industry by improving links, the UK Government has taken several steps to ensure that initial property rights for British discoveries remain in British hands. Two significant initiatives have been the establishment, by joint financing from the British Technology Group[29] and private investors, of Celltech Ltd. (1980) and Agricultural Genetics Company Ltd. (1983), companies designed to exploit research at institutes of the Medical Research Council and the Agricultural and Food Research Council, respectively. Like the US start-ups, these companies ideally should exploit tertiary research, and provide a bridge between universities and industry, which, despite other initiatives, may fail to recognise or to exploit commercially applicable research. It is interesting to note, however, that Celltech currently holds more contracts with American and Japanese firms than with firms from the UK.

Finally, there appears to be a general trend by the UK Government towards increasing support of applied research at the possible expense of fundamental research (to an extent seen as alarming by proponents of basic research). It is difficult to measure how much fundamental research underpinning biotechnology has been reduced, but it has been suggested that many areas of fundamental research relevant to biotechnology have been sheltered from severe cutbacks due to their foreseeable industrial applicability.

In *Japan*, although companies traditionally rely on universities to train researchers, scarcely any examples can be found of joint university/industry contract research. The predominant form of research links in Japan occur between industry and government research establishments (such as the Fermentation Research Institute and the National Chemical Laboratory which belong to MITI) rather than universities. Traditionally, Japanese industry has rarely if ever performed contract research at Japanese universities, probably due in large part to strict government regulations over such research agreements and legal restrictions which have, in the past, prevented industry from claiming property rights to any discovery spawned in National University Laboratories. Moreover, at the National Universities, professors are not permitted to perform consultancy services for industry, although information flows frequently from one to the other through personal contacts and informal advice.

The Bio-industry Development Centre (BIDEC), an organisation representing Japanese bio-industry and formed under the aegis of the Japanese Industrial Fermentation Association provides an important information sharing mechanism for industry, academia and government, and serves to raise awareness in Japanese industry of opportunities presented by biotechnology; however, industry/university co-operation does not take place at the research level.

Generally speaking, Japanese ministries, research institutes and universities have been rather insular and interaction has been very limited. Experts have noted that this is already creating problems in some areas where manpower shortages require closer co-operation and more manpower exchange. In order to lessen compartmentalisation of research and to encourage greater co-operation between government, industry and university R&D, measures have been recently introduced to eliminate existing legal and administrative bottlenecks and to facilitate greater research exchange.

The Advisory Board of the Ministry of Education, for example, in policy recommendations of February 1986, advocated a substantial overhaul of the university system. The report criticised the university system as being too classical, and pointed out that although the Japanese were able to rely on foreign research to catch up in the underpinning sciences of biotechnology where they were behind, reforms at home must now be sought. Closer co-operation between universities and firms in Japan must be facilitated, according to the report, and more exchange of researchers must be introduced.

The Ministry of Education has in fact been promoting more corporate participation in university research projects. Although the number of industrialists working with university research groups is still quite small, gradual improvements are evident in the increasing numbers of joint and contract research arrangements between universities and firms[30]. Two to three years ago, universities and companies began to jointly fund projects, where participants worked together in university laboratories. This coincided with a change in government policy regarding property rights resulting from university-spawned research, to their present state where patent rights are subject to negotiation between parties. In addition, more grant money is being received from industry to fund university research. Furthermore, at the level of the Science Council of the Ministry of Education, an active debate is taking place regarding the interpretation of a law which currently prohibits university employees from conducting research in company laboratories.

Finally, at some universities, consideration is being given to establishing special facilities for collaborative research in biotechnology, due to the need for new and improved equipment, among which bioreactors and gene diagnosis systems have been mentioned. Projects would be of limited duration and once a discovery was made, subsequent work would be carried out at the company facility, to avoid infringement of or conflicts over property rights.

In *Switzerland*, not unlike Japan, formal university/industry links are generally limited to consultancy services and to industrial support of basic research, and rarely entail contract arrangements. The three major industrial actors in Swiss biotechnology (Hoffman La Roche, Ciba-Geigy and Sandoz) rarely engage in contract research with Swiss universities, relying on them primarily for training and for contact with the international scientific community. Although the three Swiss multinational chemical firms appear rather satisfied with their university relations in Switzerland, relying on US and UK venture capital firms as well as their own private research institutes for developmental research agreements, the limited nature of such relations may impair the ability of smaller industrial concerns, which rely heavily on Swiss research, to exploit the academic research base.

Small companies have close relations with the Swiss Federal Institute of Technology (ETH), particularly in the area of biotechnology equipment, and limited funds are available from the Swiss applied research funding agency (CERS) to stimulate co-operative research. However, with the exception of the Federal Institute of Technology, Swiss higher education has been slow to adapt to the rapid advancements in biotechnology and as a result, may not offer internationally competitive research to small Swiss firms. One means of introducing new techniques in Switzerland has been through participation in the COST (Coopération Européenne du Recherche Scientifique and Technologique) Programme. The COST Programme has been valuable in stimulating international research contacts and most importantly in introducing new project areas. Projects relevant to biotechnology have included research on the biodegradation of lignocellulose compounds, early detection of plant viral diseases and work in plant tissue culture.

In *Belgium*, several factors have impeded the successful transfer of research results to industry. One of the more critical factors has been a mismatch between outstanding research expertise in the pharmaceutical and biomedical sciences, and industrial strength in the agro-chemical sector. However, there appears to be a growing demand in Belgium for biological research expertise in the agricultural sector. Another issue facing policy-makers has been how to contend with exploitation of excellence in Belgian molecular biological research by foreign multinationals, and the associated brain-drain.

Furthermore, national research funds tend to be spread out among fragmented groups. This occurs because of the political structure of Belgium, characterised by the existence of different regions. The fact that public moneys must support research activities carried out at eighteen universities compounds the difficulty of selectively concentrating funds. For these reasons, a strain is placed on funding bodies, making it difficult to establish excellence in emerging fields and new teams. In this last respect however, efforts are under way in the context of the "Concerted Research Actions" and the "Inter-university Attraction Poles". Initiated by the Government and managed by the Science Policy Planning Office (SPPS), a major part of these programmes covers the biotechnology sector.

Another obstacle militating against transfer stems from the industrial side: experts contend that, in general, Belgian firms need to improve their efforts and ability to absorb Belgian research. These firms have been often characterised as rather conservative, cautious or reluctant vis-à-vis research, unable to assess the commercial or industrial viability of scientific discoveries, lacking resources to afford evaluations of new areas of science and technology for potential commercial applicability, or simply not formally dealing with the need to integrate new technologies into production processes. This is an area where the Government could help by promoting greater awareness in industry, which in turn might augment limited national funds to support emerging research teams. Belgian firms also need outside information to know where and how to acquire expertise to better absorb research and apply it commercially, whether from the Belgian research base or from abroad.

The Belgian Government has in fact implemented several programmes in order to encourage greater transfer of research to industry. One was the programme for technological actions and innovations, to enhance transfer of research results between universities and industry. Under this programme, which applied to other areas as well as biotechnology, the company concerned gained property rights to research results, but had to reimburse the state if successful within five to ten years. Responsability for this programme has recently been transferred to the regions. An additional programme, launched by the SPPS and entitled "The Belgian Co-ordinated Collection of Micro-organisms", concerns amongst others, the transfer of microbial strains to Belgian industry.

Another programme, unique to biotechnology, has been established under the Institute for the Encouragement of Scientific Research in Industry and Agriculture (IRSIA), the national applied research council, whereby industry and universities are fused in precompetitive research projects at university laboratories. Under this programme, the universities and IRSIA provide financing, and industry is given a free window on research. One obvious problem which seems immediately apparent is the likelihood that without a substantial investment by a number of companies, there will be little incentive for industry to actively direct developments.

IRSIA supports programmes in monoclonal antibodies, fermentation, genetic engineering and protein synthesis. Downstream processing is acknowledged to be an under-evaluated and under-supported area of research; expertise should be developed in this area even if Belgium does not seek to become self-sufficient, but rather to import the necessary technology.

In addition to national initiatives, growing regional efforts have been made to support Belgian biotechnology. Most notable have been the establishment of two university spinoffs, Biogent and Plant Genetic Systems, with support of the Flemish Government. Both are research companies operating under contracts, and may help to exploit emerging Belgian research, if not provide a bridge to transfer results to established industry. In addition, in Wallonia, the Compagnie de Developpement Agricole and the Compagnie de Developpement des Biotechnologies have been established, partly with the help of Walloon funding.

In *Denmark*, efforts have been made by the universities, several government departments, and by the Academy of Technical Sciences (ATV) to foster greater co-operation between universities and industry. These efforts include the establishment of three science parks. Traditionally, contact between the academic and industrial communities in Denmark has been rather limited, partially due to the prevailing philosophy that research must be open and freely available. This has been the case particularly in the older universities, where until a decade ago, the occurrence of consultancies and contract research was rather uncommon. Recently, however, a general tendency towards greater co-operation has been emerging. Co-operation is particularly evident at the Technical University, the Academy of Technical Sciences (ATV) institutes, and at the newer universities.

In *Italy*, there seems to be no coherent government policy aimed at fostering industry/university links encouraging commercial application of university R&D, and bureaucratic difficulties appear to have created obstacles to large-scale co-ordinated activities between industry and universities. Interactions do take place, on an informal and limited basis, such as consultancy, retraining of industrialists at universities (which is occurring more frequently, though often abroad), and small R&D contracts from industry to universities. However, biotechnology has not created as many new dynamic interactions between the industrial and academic sectors in Italy, as it has in many other OECD countries.

One of the major impediments to improved relations between industry and academia in biotechnology is the inapplicability of much university research to industry. The National Research Council (CNR) ostensibly makes research funding more easily obtained if parallel research is carried out in industry, and periodical feasibility studies take place to determine the industrial viability of certain academic research projects. However, a technological gap between the basic research performed in universities and institutes, and the production-oriented research interests of industry, serves to inhibit the formation of dynamic university/industry links. Experts are currently considering means of co-ordinating private and public funding to establish centres for pre-competitive applied biotechnology research.

In *Canada*, the federal Government has taken several initiatives to enhance industry/university links in biotechnology. The National Research Council administers a Programme for Industry/Laboratory Projects (PILP) to encourage, through cost-shared financing, joint research projects between industry and government or university laboratories. In August 1983, PILP was extended to include a special Biotechnology programme, whereby firms will be required to contribute a minimum of 20 per cent to the total cost of a project, and such company contributions will be expected to increase with time as the industry grows commercially. In addition, NRC's biotechnology programme, which presently includes over six hundred people and an annual budget of over C$ 50 million, works to support the creation of new bio-industries and to transform existing Canadian resource industries that lack R&D expertise in biotechnology by providing financial and technological support.

The federal Government of Canada has also established R&D networks in each of the priority areas, namely forestry, human and animal health care, plant strain development, nitrogen fixation, waste treatment and mineral leaching and mining. These networks, with participation from industry, academia and government, are intended to promote communication and stimulate industrially relevant research proposals. While the communications function has been successful, the expediency of these networks to stimulate research activity is questionable, as little direction is given to their progress in achieving stated goals, and virtually no funds are provided as incentives to launch research efforts.

Industrial Research Chairs in Biotechnology have also been established as part of the federal programme, to encourage closer co-operation between universities and Canadian industry. Under this scheme, the Natural Sciences and Engineering Research Council (NSERC) and private firms jointly finance university chairs in areas relevant to the biotechnology industry. Finally, the National Research Council (NRC) operates three research institutes in Saskatoon, Ottawa and Montreal, each designed to co-operate with industry and to bridge academic and industrial research. The Plant Biotechnology Institute in Saskatoon has been particularly noteworthy in bridging co-operation between the University of Saskatchewan and several small biotechnology companies.

*
**

The issue of improving industry/university links represents an old but paramount theme in science policy. With the emergence of biotechnology, this theme has resurfaced with greater urgency, prompting several governments with long-standing weaknesses in this area to remove barriers and adopt new initiatives to facilitate the commercial exploitation of university and government research. While many of these initiatives adhere to traditional patterns, including financial support for co-operative R&D projects, exchanges of research personnel, and targeted support for industrially relevant research in universities, several developments reflect an evolved and perhaps more profound degree of concern. These developments include the establishment of R&D companies by governments, a visibly growing awareness of the intellectual property issues deriving from industry/university co-operation, and the utilisation of biotechnology, in countries where links have been almost non-existent in the past, to justify profound policy re-orientations.

In most cases, these initiatives and developments may be regarded as positive, particularly in those countries where in the past, traditions and policies have militated against successful alliances between industrial and research communities. However, a tendency in some countries to bias support towards "industrially relevant" or applied research in the universities at the expense of fundamental research, or to allow industrial research support to gradually supplant government support, could in the long run undermine the fundamental research base on which future progress in biotechnology depends.

IV. TRAINING AND MOBILISATION OF R&D MANPOWER

1. University training programmes

a) Interdisciplinarity and teamwork

The expansion of the biotechnology industry, both through the growth of biotechnology-based firms, and the continued diversification of other industrial concerns into biotechnology R&D activities, hinges on the expertise and the availability of highly qualified R&D manpower. For this reason, educators, policy makers and industrialists have all focused considerable attention on assessing the skills which constitute, and the methods which produce the calibre of R&D expertise deemed necessary for competitive strength in biotechnology.

Most experts agree that the two predominant features of manpower needs in biotechnology are their multidisciplinary nature and their high qualification profiles. Therefore, a principal challenge facing policy-makers has been to reconcile multidisciplinarity with educational curricula in such a way as to produce the highest caliber of training in biotechnology. There has been general consensus that interdisciplinary training in biotechnology should complement, rather than supplant, the training of specialists within the underpinning disciplines. In other words, the biotechnologist should be a specialist, such as a microbiologist, biochemist or process engineer, with some additional interdisciplinary training, rather than a generalist with a "hybrid" training in all aspects of biotechnology.

Furthermore, most experts agree that the ultimate goal of multidisciplinary training should be to prepare individuals for successful interaction in research teams. Hence, early exposure to a range of disciplines within biological and engineering sciences plays an important role in enabling specialists to understand the constraints and possibilities relating to these disciplines and more fundamentally in helping to familiarise them with the terminology vital for effective communication. The achievement of effective interaction between students of biological sciences and students of engineering has been a particularly critical challenge, given prevailing conditions in most OECD countries of segregation between these disciplines.

Notwithstanding the hurdles, most countries reviewed have responded to the demands of biotechnology by effecting some reorientation of traditional educational training programmes in order to accommodate greater interdisciplinarity. These adaptations range from the institution of special interdisciplinary bachelor's or master's degree programmes, to short courses at the post-graduate level which provide academics and industrialists with the opportunity to update or broaden their training.

In 1984, reports by the Office of Technology Assessment and by the Department of Commerce noted a paucity of interdisciplinary research programmes at the undergraduate and graduate levels in the *United States*. According to the US Department of Commerce, "less than a dozen training programmes (exist) in the United States which provide interdisciplinary programmes in chemical engineering and microbiology." Furthermore, a significant percentage of the students in these programmes are foreign-born, with the largest contingents from Japan, India and the People's Republic of China. The Office of Technology Assessment (OTA) also referred to communication barriers, resulting from a lack of interdisciplinary training, between classical plant breeders and plant molecular biologists, and between biologists and engineers.

In partial response to this problem, the National Science Foundation has actively promoted multidisciplinarity (and industry/university collaboration) in the research and training process, through the establishment of several engineering research centres which promote cross-disciplinary research within a co-operative university/industry context. In biotechnology, the NSF has established a process engineering centre at MIT, instituted to foster cross-disciplinary research in the areas of: genetics and molecular biology, bioreactor design and operations, product isolation and purification, and biochemical process systems engineering.

In the *Netherlands*, where the value of interdisciplinary research has been recognised and supported for quite some time, numerous opportunities exist to achieve multidisciplinary training in biotechnology, and under the Government's biotechnology programme, project support at universities is partially contingent on multidisciplinarity. The main centres of multidisciplinary biotechnology research in the Netherlands are: The Delft University of Technology, the Wageningen University of Agriculture, and the State University of Groningen. Delft essentially trains a chemical engineer with working knowledge of the biological disciplines and a rather complete background in engineering and chemistry, Groningen (considered excellent for protein engineering) trains a chemist with a working knowledge of engineering and the biological disciplines, and the education at Wageningen is based on the biological disciplines, particularly molecular biology and biochemistry, the chemistry and engineering components being minor to rudimentary.

In addition, a special biotechnology programme has been recently established at Biotechnology Delft/Leiden (BDL), a joint biotechnology centre of the Delft University of Technology and Leiden University, currently recognised as a lead centre for plant cell biotechnology and for process engineering. Biological and engineering disciplines are integrated at BDL through six major multidisciplinary projects conducted through the co-ordinated efforts of several departments, with biochemical engineering playing a central role. BDL offers a special multidisciplinary training programme in biotechnology, whereby four years are devoted to training in one or two relevant disciplines with a broad overview of biotechnology, culminating in a doctorate or engineering degree. A subsequent year and one-half, complementary to the first part, leads to a multidisciplinary diploma. Most Dutch universities plan to introduce a similar programme, based on this format, so that during the first four years, the student would major in one of the traditional underpinning sciences, e.g. chemistry or biology, and during the final year and a half, the student would take short advanced courses, conduct a feasibility study and carry out a one year multidisciplinary project.

Short courses, e.g. a one and a half year programme on fermentation technology at Wageningen and Leiden, and one on microbiology and biochemistry at the universities of Groningen, Delft, and Amsterdam are also available, and some companies offer short courses as well.

In *Sweden*, the National Board of Universities and Colleges proposed in 1983 the institution of an interdisciplinary biotechnology programme at the undergraduate level, whereby three years would be devoted to concentration in one discipline, followed by one additional year of general biotechnology studies. Based on this proposal, an interdisciplinary biotechnology programme was established at Uppsala university, under which the final year of "biotechnology studies" covers a wide range of topics such as gene technology, animal cell culturing, and downstream processing.

Other opportunities for interdisciplinary training in Sweden at the undergraduate level exist at the University of Lund technical faculty and the Royal Institute of Technology at Stockholm, the explicit purpose of these programmes being to train biochemical engineers. Both institutions provide biotechnology training on an integrated scale, offering a biotechnology degree at the undergraduate level, and courses in a wide range of areas, such as downstream processing, enzyme immobilisation, separation techniques and bioreactor design.

At Umea University as well, students specialising in experimental biology are given some cross-disciplinary training in their fourth year, which may be devoted to "biotechnology" if opted for. This year of "biotechnology", which is not required for the degree, includes an economics course, a computer course, a course dealing with cultivation in fermentors and one in separation techniques. In response to what has so far been a highly limited enrolment, (only one out of eight possible places has been filled) other departments will soon become entitled to participate.

In the United Kingdom, the Science and Engineering Research Council contends that biotechnology is not an appropriate subject for first degree courses, but should be introduced at the doctoral level. Several post-graduate programmes exist, however, offering either a diploma or an MSc in biotechnology or in bioprocess engineering. These programmes provide a broad-based exposure to biotechnology, some being particularly oriented towards business and industrial concerns. At University College in London, and Birmingham University, long-established biochemical engineering programmes have recently been modified to enhance the biotechnology training component, and full-time MSc courses in biotechnology are offered. Other programmes exist as well, including a post-graduate course in biotechnology at the University of Manchester Institute of Science and Technology (UMIST).

In addition to degree courses, a number of short courses in biotechnology are also available at the post-graduate level, for example a gene cloning course at Leicester University and one at the Polytechnic of Central London. These one- to two-week post-graduate level courses are geared to researchers from all sectors, and focus on specific aspects of novel biotechnology. Approximately 75 per cent of participants are company based, the rest being PhD students, post-doctorates and teaching staff[31]. Moreover, part-time and modular courses of longer duration exist, e.g. a part-time MSc in biotechnology at Hatfield, and a modular option for the MSc in biochemical engineering at Birmingham. Several institutions have also taken intermediate steps, by modifying first degree single honours courses in bioscience subjects to incorporate options on principles of engineering, and a number of universities have taken other initiatives to co-ordinate biotechnology research across departments.

Other opportunities for interdisciplinary training in biotechnology include a one-year post-graduate course at the Institute for Biotechnology Studies (IBS), the only university institution supported by the Department of Trade and Industry as a centre of excellence in biotechnology. This programme, primarily designed for foreigners, provides a very broad-based view of biotechnology, including 10 weeks of intensive courses on fundamental aspects of biotechnology, followed by 40 weeks of "research training" in more applied areas, and exposure to some of the broad industrial, commercial, and policy issues relevant to biotechnology, through seminars on topics such as R&D planning in biotechnology, process economics, environmental impact and hazard assessment of biotechnological products and waste. Also, in the agricultural domain, Britain's Plant Breeding Institute is considered exemplary for interdisciplinary research on plants.

British firms have expressed some criticism regarding the training of biotechnologists, implying that existing methods might be better adapted to industrial needs. For example, some British firms regard the MSc in bioprocess engineering as overspecialised, on the grounds that qualifications should consist of broad flexibility across conventional engineering, experience in bioprocess engineering, if necessary, being achievable on the job. SERC, however, contends that courses specialising in biochemical engineering, both at the under-graduate level and the MSc level produce competent well-rounded process engineers, and that a growing, albeit small demand for specialised biochemical engineering graduates does exist. Furthermore, British firms seem to feel that there is an overprovision of MSc's in biotechnology on the whole, and that the quality of "enhanced" or "converted" graduates undertaking post-graduate degree courses may not be adequate, particularly when the conversion aims to mold engineers out of bioscientists.

Currently, a generally restrictive policy exists in British firms regarding advanced training for their employees, extramural training usually confined to intensive/short courses which last from one or two days to one week. This may be directly attributable to the generally perceived need to reduce the duration of employee absences. Nevertheless, SERC has been striving to identify ways to improve biotechnology training opportunities for industry, suggesting, for example, that perhaps not enough commercial and business content exists in the post-graduate training of biotechnologists, or that perhaps a need exists for courses and seminars to brief management on the technical/commercial implications of biotechnology.

In *Germany*, where traditional barriers between biological and engineering faculties have been relatively strong, the need for government initiatives to encourage interdisciplinary co-operation through "teamwork", and the exposure of individuals to more than one discipline, was stressed in recommendations to the Government in 1982 by DECHEMA. The particular suggestion that scientists be trained in one field intensively at the under-graduate level, and in another discipline at the graduate level, was eventually adopted by the BMFT, which has earmarked funds to encourage students to study a field outside of their specialisation at the graduate level. Under the "research stipend for biotechnology" programme, which has been allocated DM 20 million for a four year period, the BMFT will promote greater interdisciplinarity in the field of biotechnology by sending graduate students for one to two years to a research institute specialising in an area different from his area of expertise.

Unlike the United States, the United Kingdom and the Netherlands, comprehensive post-graduate degree courses in biotechnology are not available in Germany. Experts have opposed establishment of post-graduate courses in biotechnology, contending that because Germany does not have a two-tiered educational system, as exists in the US and the UK, but rather a 5-6 year coursework leading to a single academic degree, students might be considered too old for the labour market, after having completed a regular course in chemistry or biology, a post-graduate course in biotechnology, and eventually a doctoral thesis in the field. Another element of the debate has been whether biotechnology "specialisation" should be achieved in academia after graduation in one of the basic disciplines, or whether instead it would be more practical to specialise in industry. However, the majority opinion holds that rather than establishing a new biotechnology coursework, greater flexibility within institutions allowing for more interdisciplinary exposure should be pursued, and biotechnology "centres of gravity" should be established.

One comprehensive biotechnology training programme which focuses notably on the areas of microbiology, fermentation technology and biochemistry, does exist at the University of Berlin, the only academic institution in Germany offering such a programme. The curriculum entails four semesters of fundamental studies, followed by four semesters devoted 40 per cent to the study of fermentation techniques, microbiology, biochemistry and economics, 50 per cent to additional seminars, courses and practical work on the biotechnologies, and 10 per cent to the study of classical fermentation techniques. The programme culminates in the preparation of a "Diplomarbeit", report of four months experimental research work followed by an oral exam and acquisition of the title "Diplom-Ingenieur".

Short courses of varying durations and at various levels are offered by the German research institutes and by DECHEMA. Some of these courses are organised in collaboration with the Swiss Federal Polytechnic School in Zurich. In addition, the federal Länder and the universities are endeavouring to establish biotechnology courses at the German technical universities, and have integrated a biotechnology component into biology and chemistry courses at a number of non-technical universities. Finally, some companies, e.g. Hœchst AG and Bayer AG offer training for high level laboratory technicians in biology, biochemistry, and fermentation technology.

In the agricultural sector, experts have noted an insufficiency of training in basic sciences and in biotechnological techniques in the agricultural colleges in Germany. It has been suggested that more flexible combinations of basic and applied training might be necessary, and that a capability for teamwork should be improved. Training currently does not address the need to integrate various disciplines, by bringing students with varying expertise together to tackle complex interdisciplinary problems; however, some trainee courses have been set up at the research centres of the Federal Ministry for Agriculture in advanced areas of monoclonal antibody production, and plant breeding techniques.

In *Japan*, the Ministry of Education has recently made efforts to reform what has been categorised as overly "classical" education, acknowledging the need to reorganise the system in order to adapt to changing demands and objectives. As part of these efforts, the training and retraining of researchers and technicians at research centres has received substantial attention. In 1986, the ministry held two training workshops, one at Tsukuba and the other at the research institute of basic biology at Okazaki. These workshops, each one budgeted at Y 11 million, were open to company employees, one addressing itself to junior researchers, the other to senior researchers. The focus of the Tsukuba workshop was plant cell treatment, e.g. isolation, culture, cultivation and fusions, while that of the Okazaki workshop was animal development, e.g. construction of hybrid eggs by DNA injection and cell transplantations. In addition, in 1986 and 1987, a substantial amount of new or expanded coursework in modern biotechnology was introduced in Japanese universities.

While the Ministry of Education has clearly taken important steps to update training opportunities for biotechnology within its administration, the need to facilitate interdisciplinary research at the training level, which has traditionally not existed due to prevailing separatism between basic and applied science departments in the university system, has also been partially addressed. Collaborative research centres open to the staff and students of biological science departments have been recently established in ten universities, e.g. the Molecular Genetics Research Laboratory at Tokyo University, and the Cell Engineering Research Centre at Osaka University.

Other training opportunities in biotechnology in Japan are offered by the Bio-Industry Development Center (BIDEC), which sponsors a large number of short courses of direct relevance to biotechnology. In 1984, these courses in "human resource development" included lecture and discussion sessions on topics such as recombinant DNA technology, bioreactors and bioelectronics, plant and animal cell engineering, corporate strategy for biotechnology, animal cell culture and plant tissue culture. Practical training courses were also offered in areas such as bioreactors and biosensors, basic technology relating to the serum-free culture of animal cells, identification of rDNA by use of an electron microscope, technology for separation and purification, and use of computers in biotechnology. These bio-industry workshops and seminars have attracted a large and rapidly increasing number of participants rising in man-days from 578 in 1983, to 1 907 in 1984, to 2 086 in 1985.

In the Japanese agricultural sector, biotechnology training in the government research establishment has been described by experts as "very weak". A rigid hierarchical tradition prevails in government laboratories, where younger staff members are trained by older members, and opportunities to gain university training in advanced, new techniques are highly limited. In order to overcome the rigidity built into the government laboratory system and to introduce researchers with knowledge and experience in the new techniques which are lacking, the Ministry of Agriculture, Forestry and Fisheries has been accepting more post-doctoral fellows than usual. However, no efforts are apparent to develop new training curricula and, as with the Ministry of Education, no significant efforts are evident to achieve greater multidisciplinarity and to promote more teamwork.

Although in most countries reviewed, historic divisions between faculties of biological science and faculties of engineering have presented obstacles to interdisciplinary co-operation, these obstacles have been compounded in *France* by the physical separation of the Grandes Ecoles, which train the most qualified students of engineering, and the universities, which educate the post promising research scientists. Currently some engineering schools in France offer one semester of general biotechnology studies; however, although these studies provide engineers with a broad exposure to the vocabulary of biotechnology, no opportunities for research training are offered.

In order to promote greater interaction between French engineering faculties, biological science research faculties, and industry, the French Government has instituted a system entitled "Firtech", under which designated centres of excellence will act as nuclei for co-operative research efforts. "Firtech" centres have already been established at the Universities of Toulouse and Compiègne, and an additional centre will be established at Nancy. It remains to be seen to what extent the logistical impediments introduced by the physical separation of engineering and research faculties will be overcome by the "Firtech" plan.

The need for retraining in France, particularly in the agri-food and chemical industries, has been regarded as acute. However, despite a sufficient availability of retraining programmes and individual courses, as well as some legal pressure imposed on companies to encourage retraining, resistance to retraining has been observed as a widespread social phenomenon in France.

In *Italy*, experts point to inadequacies in biotechnology training, which stem not exclusively from lack of interdisciplinarity, but more fundamentally from the lack of research training in general. This problem is manifest in the fact that the "Laurea in Biologia", the Italian counterpart to the master's degree, offers no research training even in the best universities. A recent initiative hoped to remedy this problem and consequently improve biotechnology training has been the institution of the "Dottorato di ricerca" (research doctorate); however, the number of young laureates admitted to some of the programmes in genetics and biology appears to be in decline.

Other significant initiatives to improve biotechnology training in Italy include plans to institute a two to three year post-graduate course in advanced biotechnology implementation in 1987 at the University of Milan, and the institution of a school of genetics at the University of Pavia. This school, close to the University's Department of Genetics, and run in collaboration with the Institute of Biochemical and Evolutional Genetics of the CNR, offers a three year training programme, addressed to young laureates, consisting of theoretical courses and participation in research projects. Industrial scholarships are often provided to students in this programme by companies interested in biotechnology, who then hire these students upon graduation.

Finally, recommendations in the June 1986 report of the National Committee for Biotechnologies, with respect to educational and training curricula for biotechnologists, include creating a "dottorato di ricerca" PhD in biotechnology, creating postdoctoral "specialisation courses", and revising the current content of several university courses, e.g. biology, chemistry, pharmacology and agronomy to update education in biotechnology.

In *Switzerland*, no formal curricula for integrated training in biotechnology exists; however, interdisciplinary research may be conducted at the Biozentrum, a research centre connected with the University of Basel where efforts concentrate on genetics, microbiology, cell biology and biophysics, or at the Federal Institute of Technology in Zurich (ETH). Even at these institutions, opportunities for interdisciplinary training depend upon the initiative of students, and interdisciplinary teamwork occurs very rarely. Swiss universities, on the other hand, offer no opportunities for interdisciplinarity and departments are highly self-contained and not interactive, despite some efforts made to improve this situation, e.g. the organisation of retraining courses in the field of environmental biotechnology. Finally, Swiss authorities express a common concern that on the average, students in Swiss elementary schools, high schools and universities gain an inadequate degree of exposure to the biological sciences, and that this may ultimately have a detrimental effect on the ability of the public to gain a balanced understanding of biotechnology.

In *Canada*, at the under-graduate level, those universities which provide a biotechnology option generally offer fundamental coursework such as biochemistry, biology, chemistry and genetic engineering during the first two years. During the third and fourth year, the students complete their basic training in the fundamental area chosen, and begin the "biotechnology option". McGill University in Montreal has been cited as particularly strong for training and interdisciplinary research collaboration, especially in the field of medicine.

At the post-graduate level, programmes are available from science faculties (biology, microbiology, biochemistry) and engineering faculties. A recent study showed that fifteen engineering programmes offer a specialisation in biotechnology in Canada; seven for a master's in engineering, and eight for a PhD.

In *Belgium*, where a long and very rich educational tradition in the biosciences exists, biotechnology training is carried out primarily at six universities in Liege, Ghent, Brussels (two), Louvain and Louvain-la-Neuve. Two of these universities offer a master's programme in biotechnology.

In *Australia*, short courses are available at the post-graduate level for the biotechnology diploma, and research programmes exist in biotechnology leading to the masters and doctoral awards. At the undergraduate level, within the traditional framework of majors, some exposure to the broad-based scope of biotechnology is provided.

In *Finland*, only one university offers engineering training in biotechnology (Helsinki University of Technology). However, various opportunities for multidisciplinary research including molecular biology and process engineering may be found at the Technical Research Centre and the Biotechnical Laboratory at Helsinki. In addition, an interdisciplinary training programme at the graduate level, open to students from all universities and to industrialists, has recently begun at the rDNA laboratory in Helsinki. This programme currently focuses on the biosciences, but may be developed to include bio-engineering through co-operation with the Helsinki University of Technology. In *Denmark*, multidisciplinary training programmes have not been created, although some biotechnology coursework is available at the University of Aarhus, the University of Copenhagen, and the Danish Technical University. The Danish Government Programme on Biotechnology has identified a general need to modernise and increase the biological content of education in Denmark, but has not yet specified at what level or how, nor clearly evaluated the adjustments which may be necessary to improve biotechnology training curricula.

b) Industrial exposure

In addition to the emphasis placed on interdisciplinarity, the exposure to practical industrial problems as a part of the training process has also become recognised as an expedient means of facilitating rapid integration of R&D manpower into industrial biotechnology. In most countries reviewed, visible efforts have been made to increase the contacts between universities and industry at the training level in biotechnology. It appears, furthermore, that a correlation exists between the willingness or ability of individuals and institutions to adapt to multidisciplinarity, and the willingness or ability to foster industrial contacts, as those countries with relative successes in the former have also demonstrated some successes in the latter.

In the *United States*, where a highly flexible university system has already encouraged a proliferation of university/industry links which expose students to industrial research problems at the training level, the MIT Process Engineering Center, where students are exposed to and perform fundamental research on engineering problems of significance to industry, represents an additional opportunity for industrial contacts at the training level. Moreover, university programmes for technician training in areas relevant to biotechnology are often run in close collaboration with neighboring firms.

In the *Netherlands*, exposure to industrial problems at the training level occurs at Biotechnology Delft/Leiden, where all projects are performed in co-operation with industry. In addition, a new programme at Delft/Leiden, entitled the "Industry Oriented Training Program in Biotechnology" has been proposed as an alternative to the more research-oriented programmes offered at other universities. According to the proposal, the programme would offer a combination of theory and practice at an estimated cost per course of Gld 30 000-40 000 (financing expected from different sources including the EEC and industry). Business-oriented courses would be included, e.g. a course on patent protection, and a biotechnology diploma would be granted upon successful completion of the programme.

A policy which has been adopted in *Sweden*, in addition to the "Adjunct Professor" scheme mentioned earlier, whereby industrialists spend approximately one day per week teaching at universities, is the allowance of graduate researchers pursuing the PhD to be partially employed by industry, provided that their employment relates directly to their thesis.

In the *United Kingdom*, several programmes exist which are designed to promote contact with industry as an integral part of the training process and which are applicable to biotechnology. At the undergraduate level, several universities and polytechnics offer "sandwich" courses which promote contact with industry by allowing students to spend either one year, or several periods of a few months as interns in an industrial environment. Due to economic difficulties, however, placement is sometimes difficult[32].

At the post-graduate level, CASE "Co-operative Awards in Science and Engineering", granted by SERC, provide partial funding for a three year doctoral research programme based on a joint research project between an academic department and a company. According to this scheme, the company contributes to the financing of the project, the student spends part-time working on this project in the firm, and any resulting property rights belong to that firm. Thirty "CASE" awards per annum exist for biotechnology, and an increase is intended.

At the post-doctoral level, the "Royal Society/SERC Industrial Fellowships" programme permits university scientists to work on a project in an industrial environment, and conversely, permits industrial engineers to take courses or conduct research projects at the university.

SERC's "Teaching Company Directorate", co-funded by the Department of Trade and

Industry, provides yet another scheme for industrial contact as an integral part of training, whereby one company co-operates with a university department to conduct a research and training programme. Generally, the research is carried out by university staff and associates in the firm, sometimes on university premises. Out of 150 schemes in SERC, 5 currently exist in biotechnology, and experts point to a growing interest amongst companies to participate in the programme, which offers a special opportunity for the training of bioscience R&D staff and bioprocess engineers.

In *Germany*, the University of Berlin's biotechnology training programme includes 26 weeks of training in an industrial enterprise, and at the Society for Biotechnology Research in Braunschweig (GBF), plans for an industrially sponsored training programme may be implemented.

In *Canada*, undergraduate programmes in biotechnology generally do not include industrial internships. However, there appears to be an increasing awareness in the universities of emerging industrial needs and deficiencies in traditional training. This has apparently resulted in a greater focus on practical, production-oriented research, which was formerly lacking. Moreover, the juncture of the universities of Guelph and Waterloo to form the Guelph Waterloo Biotechnology Institute emphasises research with industrial applicability, and is managed by an industrial board of governors.

In *Belgium*, the Science Policy Planning Office runs an industrial "stage" programme, notably in biotechnology, which supports young researchers in industrial training programmes after two years of university training, in order to help facilitate more rapid integration of manpower into industrial practices. This programme will come to an end in 1987.

2. Quantifying and addressing biotechnology manpower needs

In addition to the issue of establishing appropriate curricula, many OECD governments have addressed the need to attract greater numbers of students and faculty to biotechnology-related disciplines. Although available estimates of overall demand for biotechnology manpower for the next five to ten years seem rather modest[33], apparent shortages of skilled personnel in all countries reviewed have already surfaced as potentially critical long-term bottlenecks to biotechnology development.

Several governments have addressed these existing or anticipated shortages through efforts to recruit more and younger scientists to various fields through new faculty appointments, general enhancement of financial opportunities for academic scientists, etc. Some of these efforts to remedy shortages have not yet had a perceptible impact, because of an inevitable time lag before educational and training adjustments may be felt in the labour market. Some countries are also pursuing policies to combat the incessant problem of brain drain which clearly exacerbates manpower shortages.

Although there appears to be an adequate supply of personnel trained in basic biosciences in the *United States*, shortages of manpower relevant to biotechnology have been cited as "acute" in biochemical engineering, particularly chemical engineers with backgrounds in biochemistry and microbiology who can design and develop the production technologies required for large-scale manufacturing[34]. In 1984, The Office of Technology Assessment predicted that shortages in both bioprocess engineering and industrial microbiology in the United States would become particularly acute when companies involved in biotechnology R&D began scale-up and production activities. Another serious shortage area pinpointed in 1984 and recently the focus of attention is plant molecular biology, where a shortage of professors currently exists, and where a serious drain of PhDs to industry has been anticipated, particularly with the growing industrial interest in plant biotechnology. The NSF has recently targeted plant biotechnology in an effort to respond to this need, hoping to train hundreds of post-doctoral researchers in interdisciplinary plant research[35].

In the *United Kingdom*, despite the wide availability of courses in biotechnology, or containing a biotechnology training element, experts have noted an insufficiency of effectively trained persons in several key areas. Although plenty of highly qualified students may be found in bioscience departments overall, shortages exist in departments of microbial physiology and plant cell biology, which, according to some accounts, are considered unpopular with high calibre students[36].

The Spinks Report (1980), and the Royal Society Report (1981), both called attention to training requirements in new areas of molecular biology, as well as the need for more training in microbial physiology, fermentation, bioreactor design and downstream processing. Since then, developments have occurred in molecular biology (e.g. rDNA, gene sequencing, DNA synthesis and monoclonal antibodies), but comparable advances in microbial physiology, biochemistry or fermentation technology are not readily apparent. In addition to anticipated shortages of experienced bioprocess engineers, demand for bioscientists with team management experience is also expected to exceed supply. A brain-drain of competent bioprocess engineers out of the UK has been cited as a particularly serious factor, attributed to non-competitive salaries in British firms and relatively few posts, which may aggravate these shortages. The Government, however, has been pursuing several measures to counteract the brain-drain[37].

SERC research priorities are in part based on critical shortages of manpower; thus, as stated earlier, (Chapter II, 1) SERC research studentships are often provided in areas requiring increased numbers of personnel. For example, until recently, no training expertise existed in scale-up of mammalian cell cultures; therefore, research grants and studentships were provided to encourage more work in this area. In addition, the University Grants Committee has also taken an initiative to strengthen research centres in biotechnology through the provision of new permanent posts.

In *France*, where a lack of competitive expertise in genetic engineering was viewed in 1982 as a critical impediment to the development of biotechnology, consequent efforts to promote this area succeeded in producing a currently adequate, if not excessive, supply of well-trained genetic engineers. However, shortages of manpower with combined expertise in genetic engineering and other disciplines, e.g. immunology, have persisted, suggesting an enduring lack of effective interdisciplinary training and a previous underestimation of the need for genetic engineers with additional expertise.

In contrast with the earlier focus on genetic engineering as the primary limiting factor in French biotechnology, experts involved in the Programme Mobilisateur now point to an insufficient knowledge of protein biochemistry[38], and lack of competitive expertise in purification techniques. Some experts also anticipate a long-term need for greater expertise in the interface of biology and information technology.

Sweden has designed several policies to facilitate manpower mobility, and to recruit younger and more researchers into the field of biotechnology, as the lack of sufficient numbers of qualified research personnel appears to be the single most important hindrance to the continued development of Swedish biotechnology. In 1983, the Swedish Board for Universities and Colleges recommended an increase in the number of professorships and post-doctoral positions in biotechnology (as well as support for equipment and chemicals, and the strengthening of Swedish plant research). The resulting legislation provided Skr 12 million over five years for 12 new professorships, and 40 new post-doctoral positions, the latter designed to enable universities to recruit younger scientists.

However, in spite of the greater availability of positions offered by the Government, academics still complain that the demand for bright students in the biological sciences is much higher than the numbers of students enrolling in programmes and that too few bright students are attracted to scientific research careers. Amongst those who are, most choose industry over academia (this is much more of a problem with graduates of engineering faculties than graduates of biological faculties), quite evidently because of the notable difference in salary levels.

The Government has attempted to overcome this problem by making academic research more flexible and financially attractive. Thus, a statute was changed such that university professors are legally entitled to conduct research or accept consultancy contracts for industry, or even create their own university spin-off companies[39]. However, as noted earlier, it remains unclear for the moment whether the proliferation of small start-up firms in Swedish biotechnology will mitigate the drain of researchers into industry, thereby preserving the fundamental research base, or whether it will aggravate the risks to industrial development by creating a competitive, rather than a co-operative climate between academics and established industry.

In regions where a disparity between areas of scientific excellence and the needs of established local industry has been a contributing factor to manpower drain (such as in the region of the northern city of Umea), the small company start-ups may provide greater incentives for academics to remain. One expert also suggested that more dynamic links should be promoted between fundamental science and established industry in the region. For example, only three people are working on plant biotechnology at Umea. Given the industrial structure of the region which is based largely on forestry, plant biotechnology might represent an appropriate field for Umea University to develop and expand.

Manpower weaknesses in Sweden are perceived in cell biology and animal cell culturing, molecular plant genetics, classical microbiology and yeast cell genetics. Shortages of competent scientists with an integrated training in the various disciplines underpinning biotechnology have been noted, as well as a need for business skills amongst small biotechnology entrepreneurs.

Denmark suffers from a generally low concentration of manpower in most areas of academic research underpinning biotechnology. The Danish Government Programme will attempt to address this issue by allocating approximately 80 per cent of its funds to the strengthening of existing research groups, and to the establishment of research networks. Under the network system, the receipt of research grants will be contingent upon a commitment by research units to collaborate with other research groups. It is hoped that by providing such an incentive for co-operation, Danish research groups will begin to generate the critical masses necessary for stronger and more competitive research.

In *Switzerland*, the academic research system has, according to several experts, been responding too slowly to the needs of industry. Ideological resistance to new technologies has been reinforced by slow infiltration of "new blood" at universities. This has been attributed to guaranteed tenure and in some cases, to absolute restrictions on the total number of employees at educational institutions regardless of those institutions' expansionary needs. Although research efforts have been extremely strong in several areas including molecular biology and genetic engineering, critical masses are not often achieved. Notable weaknesses in Swiss public sector research may be found in plant genetic manipulation and large scale cell culture.

In the *Netherlands*, the demand for industrial manpower in biotechnology is already quite strong and shortages of competent people in general have already been felt, evident in the fact, according to one expert, that industry will often accept highly qualified university scientists as consultants, regardless of whether they simultaneously work for other companies.

In *Germany*, experts have indicated that approximately 300 biochemical engineering graduates can be placed each year, and that this number would be unlikely to increase drastically, although emphasis has been placed on the increasing needs of the food industry. Generally speaking, the supply of personnel in areas related to scale-up and bioprocess engineering seems to be adequate.

In *Japan*, biotechnology manpower shortages do not appear to pose serious problems in the private sector. Expertise in bioprocess engineering and industrial microbiology has been traditionally plentiful, and large companies have managed to gather sufficient numbers of highly skilled R&D personnel in the newer areas of biotechnology such as rDNA and hybridoma technology, through hiring Japanese researchers from abroad, sending employees overseas for training, and some in-house instruction. However, smaller companies which cannot afford some of these options, particularly that of training researchers abroad, are at a distinct disadvantage. In certain areas of the public sector, biotechnology manpower shortages are acknowledged to be a troublesome problem, aggravated by a lack of manpower mobility and obstacles to creating critical masses and teams. In hospital research laboratories, experts note a serious shortage of technicians, particularly in biotechnology research.

Although the general level of *Italian* biologists has been considered quite good according to international standards, the number of scientists qualified in advanced biotechnology remains quite limited, relative to national demand. Recommendations in the June 1986 report of the National Committee for Biotechnologies, with respect to meeting manpower requirements for Italian biotechnology, include doubling the number of scientists within five years in universities and public research organisations, and establishing special grant money for young graduates willing to carry out research in biotechnology.

According to a study commissioned by the National Research Council of *Canada* on Human Resources for Biotechnology, lack of skilled manpower represents a critical problem in Canada, perhaps the most serious impediment to establishing a strong industrially-oriented R&D base in biotechnology. The conclusions of this report, published in May 1984, stated that it was not obvious that the broadly trained students needed for industrial biotechnology were or would be coming out of the university system with adequate training or in sufficient numbers to satisfy demand. Other experts, however, have suggested that industrial demand for highly qualified research personnel is growing at a slow enough rate that current supplies of personnel should be adequate for the present time. In 1984, the NRC Task Force estimated the demand for doctoral graduates in biotechnology-related disciplines at 600 to 800 over the next five years.

Amongst the schemes instituted by the Natural Sciences and Engineering Research Council (NSERC) to augment biotechnology training opportunities are an Industrial Fellowships Programme in Biotechnology, intended for doctoral graduates seeking industrial employment in Canada, and an Industrial Professorships scheme, through which eight professorships have been created in biotechnology.

Included amongst the more critical shortage areas cited for Canada are basic plant biochemistry and biochemical engineering. In the latter area, the number of programmes as well as faculty and students is very small relative to other strategic sectors in Canada, e.g. micro-electronics, computers and robotics, and concern has been expressed over the perceived shortage of production-oriented biochemical engineers. In general, critical masses are lacking due to a dispersal of manpower, and a need exists for people with training in production and marketing, and for senior management personnel with technical backgrounds appropriate for biotechnology. Canada is weak in fermentation technology, downstream processing and separation technology. With regards to the brain-drain of competent scientists out of Canada to the United States, experts have suggested that efforts could be made to encourage Canadian repatriation and generate more enthusiasm through more clearly stated national goals in biotechnology.

According to a study made by the Academy of *Finland*[40], a lack of skilled manpower presents a critical problem in several areas of Finnish biotechnology. This problem may be addressed by a new proposal for research training which would seek to double the number of qualified researchers in Finland during the next five years.

*
**

Along with the need for university/industry links, the general importance of interdisciplinary training and industrial exposure has been an old theme in science policy, recognised and discussed at policy levels long before the emergence of biotechnology. In the context of biotechnology, not surprisingly, general consensus was arrived at quite early that university/industry links and interdisciplinarity would be critical to its development. A large number of initiatives reorienting educational institutions to the interdisciplinary training requirements of biotechnology and fostering greater industrial exposure in this field attest to the seriousness with which governments have responded to this need.

However, in terms of channeling sufficient quantities of manpower into those areas deemed necessary for underpinning industrial development, critical shortages have persisted in all countries reviewed in at least two areas, these areas covering a fairly wide range of categories of expertise. Some countries appear to have had substantially greater success than others in terms of producing adequate numbers of highly skilled manpower at home (Germany, the US), while others have tapped a combination of expertise at home with expertise abroad (Japan) to compensate for research weaknesses. Nevertheless, the fact remains that manpower shortages and problems have persisted in all countries; hence, continued attention by governments to ameliorate biotechnology manpower conditions would seem highly appropriate.

Finally, for many OECD countries, the need to increase the general exposure of students to the fundamental and applied aspects of biological science is imperative. Only by educating the public to understand the science beneath and the ramifications of biotechnology, will public acceptance and support of biotechnology be ensured.

NOTES AND REFERENCES

1. Rip, A. and Nederhof, A., "Between Dirigism and Laissez-Faire: Effects of Implementing the Science Policy Priority for Biotechnology in the Netherlands", *Research Policy 15*, 1986.

2. Science Council of Canada, *Biotechnology in Canada: Promises and Concerns*, Proceedings of a Workshop Sponsored by The Institute for Research on Public Policy and the Science Council of Canada, September 1980, and Ministry of State for Science and Technology, *Biotechnology: A Development Plan for Canada*, Report of the Task Force on Biotechnology to the Minister of State for Science and Technology, February 1981.

3. It has been suggested that DRIE reconcile its dual mandate of industrial and regional development in terms of biotechnology, identify specific priority areas for industry, and where congruent with the national programme, co-ordinate a tangible research and development strategy with the other actors, particularly the NRC. *Seeds of Renewal: Biotechnology and Canada's Resource Industries*, Science Council of Canada, Ottawa, 1985.

4. Canadian educational policy falls under provincial, rather than federal responsibility, and federal influence is therefore limited to grants. Currently, NSERC provides approximately C\$4.5 million per annum through strategic grants to support industrially applicable research in biotechnology.

5. Comitato Nazionale Per le Biotecnologie, Ufficio del Ministro per il Coordinamento della Ricerca Scientifica e Tecnologica, *1 Rapporto*, 1986.

6. The definition of the term biotechnology used here encompasses process engineering, cell-tissue cultivation, protoplast fusion and genetic engineering techniques.

7. US Government Interagency Report (1983, Table B-14a).

8. Yanchinski, S., "Boom and Bust in the Bio Business", *New Scientist*, 22nd January, 1987.

9. Particularly the discovery of and subsequent application of natural enzymes in chemical and pharmaceutical processes, immobilisation techniques, continuous fermentation techniques and advancements in bioreactor design.

10. IBS is based on University College, London University, the University of Kent, and Polytechnic of Central London.

11. US Government Interagency Report (1983, Table B-14a).

12. Webber, D., "Enterprising Biotechnology Firm Directs Research at Food Products", *Chemical and Engineering News*, 19 August 1985, and Bylinsky, G., "Test Tube Plants Hit Pay Dirt", *Fortune*, 2nd September, 1985.

13. Kalter, R., "The New Biotech Agriculture: Unforeseen Economic Consequences", *Issues in Science and Technology*, Fall 1985.

14. US Congress, Office of Technology Assessment, *Technology, Public Policy, and the Changing Structure of American Agriculture*, OTA-F-285 (Washington, DC: US Government Printing Office, March 1986).

15. Dunnill, P. and Rudd, M., *Biotechnology and British Industry: A Report to the Biotechnology Directorate of the Science and Engineering Research Council*, London, 1984, p. 21.

16. Estimates that development of atrazine-resistant soybeans will allow for double the use of the herbicide on corn, producing tens of millions of dollars in additional sales. See: Meier, B., "Technology: A Special Summary and Forecast of Scientific Developments Affecting Business", *Wall Street Journal*, 3rd September, 1985.

17. *Ibid.*

18. Beardsley, T., "Plant Genes Now Fashionable", *Nature*, Vol. 324, 13th November, 1986.

19. *Seeds of Renewal, op. cit.*, p. 31.

20. Concertation Unit for Biotechnology in Europe, Commission of the European Communities, *Towards a Market Driven Agriculture*, Consultation Document, December 1985.

21. Krassner, M., "Finding Parents for Orphan Drugs", *Chemical and Engineering News*, 26th August, 1985.

22. US Congress, Office of Technology Assessment, *Human Gene Therapy — A Background Paper*, (December 1984); US Congress, Office of Technology Assessment, *Impacts of Applied Genetics*, (April 1981); US Congress, Office of Technology Assessment, *The Role of Genetic Testing in the Prevention of Occupational Disease*, (April 1983).

23. United States Environmental Protection Agency, *The Proceedings of the US EPA Workshop on Biotechnology and Pollution Control*, Washington, D.C., 10th November, 1986.

24. Blumenthal, D., Gluck, M., Louis, K. and Wise, D., "Industrial Support of University Research in Biotechnology", *Science*, vol. 231, 17th January, 1986.

25. Organisation for Economic Co-Operation and Development, *Industry and University: New Forms of Co-operation and Communication*, Paris 1984.

26. US Congress, Office of Technology Assessment, *Commercial Biotechnology: An International Analysis*, (Washington, DC., January 1984), Chapter 12.

27. O'Sullivan, D., "West German Chemicals Make Rapid Gains", *Chemical and Engineering News*, 3rd February, 1986.

28. *The Netherlands: Busy in Biotechnology*, Proceedings of a Symposium, 29th May, 1985, Delft, p. 22.

29. The National Research Development Corporation (NRDC) and the National Enterprise Board (NEB) are statutory bodies which operate together since 20th July, 1981 under the name of the British Technology Group (BTG), the role of which is to concentrate on the translation into commercial products of new research ideas.

30. Ministry of Education, Science and Culture (Monbusho), *Research Co-operation Between Universities and Industry: Ongoing Creative and Advanced Research for the Future of Japan*, (Provisional), Japan, 1986.

31. Pearson, R., and Parsons D., *Enabling Manpower for Biotechnology in the UK*, a report prepared for the Biotechnology Directorate of the Science and Engineering Research Council by the Institute of Manpower Studies, December 1983, p. 48.

32. Chopplet, M., Centre d'Etudes des Systèmes et des Technologies Avancées, *Les biotechnologies dans le monde*, Paris, 1985, p. 17.

33. The Royal Society, *Biotechnology and Education*, London, 1981.

34. US Department of Commerce, International Trade Administration, *Biotechnology*, July 1984.

35. Crawford, M., "OSTP Ponders Plant Research Initiatives", *Science*, vol. 231, 17th January, 1986, p. 212.

36. Pearson, R., and Parsons D., *op. cit.*, p. 43.

37. U.S. Congress, Office of Technology Assessment, *Commercial Biotechnology: An International Analysis*, (Washington, DC., January 1984), p. 338; and Pearson, R., and Parsons D., *The Biotechnology Brain Drain*, a report prepared for the Biotechnology Directorate of the Science and Engineering Research Council by the Institute of Manpower Studies, December 1983.

38. Yanchinski, S., *op. cit.*

39. This was probably a direct response to the drain of manpower from universities to industry. However, in actuality, most of these activites took place on a large scale before the legislation was enacted.

40. The Academy of Finland, *Programme for the Development of Biotechnology and Molecular Biology for the Years 1988-1992*, (Helsinki 1987).

Part II

CANADA-OECD JOINT WORKSHOP ON NATIONAL POLICIES AND PRIORITIES IN BIOTECHNOLOGY,

Toronto, Canada
7th-10th April, 1987

I. REPORT OF THE WORKSHOP

This workshop brought together about 70 experts and policy-makers (see list Annex I), many of them with senior responsibilities in the field of biotechnology, from Belgium, Canada, Denmark, Finland, France, Germany, Italy, Japan, the Netherlands, Norway, Portugal, Sweden, Switzerland, the United Kingdom, the United States, Yugoslavia and the EEC. Also present were a number of industrialists invited by the OECD's Business and Industry Advisory Committee (BIAC). It was chaired by Dr. John R. Evans, Chairman of Allelix Inc., Ontario, Canada.

In their opening remarks, officials from Canada outlined the considerable progress that biotechnology has made over the past decade and stressed the importance of this Workshop as the first opportunity for OECD Member countries to review their biotechnology policies together. The OECD representative placed the Workshop into the larger perspective of past and ongoing CSTP-work on biotechnology. The Workshop Chairman, Dr. John Evans, Chairman of Canada's National Biotechnology Advisory Committee, stressed that the goal of the Workshop was not necessarily to reach consensus on all problems, but to define the most topical issues presently confronting policy-makers, including those where there was lack of agreement between countries and where further work could help to reconcile differing viewpoints and approaches.

During the Workshop, and particularly during the discussions on policy recommendations, it became clear that participants overwhelmingly agreed on the policy issues which were at present most urgent in the field of biotechnology. There was also a large measure of agreement on the roles which governments should play in responding to these issues, even if some national policy differences, mainly regarding support for industrial R&D, remained. Thus, the Toronto Workshop and its recommendations allow for a first international answer to one of the more vexing questions which have accompanied the progress of biotechnology; namely, the question of the appropriate role of governments as compared to the roles of industry and academia in the development of this new technology.

The Workshop was divided into two parts; keynote presentations and discussions on the first day, and three working group Sessions leading to draft recommendations discussion on the second day.

Keynote presentations and discussions were grouped into five different sessions:

1. National policies and programmes in biotechnology;
2. Setting strategic priorities — the Japanese system;
3. The commercialisation of government and university research, and public acceptance of biotechnology;
4. Training and mobilisation of human resources and the flow of scientific information; and
5. Safety and regulations, R&D and international co-operation in biotechnology.

The following are summaries of the keynote presentations and discussions. The texts of the keynote presentations can be found in the Annex.

1. National policies and programmes in biotechnology

Dr. Ronald Coleman, Government Chemist, Chairman, Interdepartmental Committee on Biotechnology, Department of Trade and Industry, United Kingdom, gave a keynote presentation on this subject, providing an overview of the biotechnology programmes and activities in major OECD countries. He identified four key factors influencing development: the economic and social climate; public support for basic science research; the supply of trained human resources; and the identification of national industrial targets.

The first factor — economic and social climate — affects the development of high-technology industries in general and is not specific to biotechnology. This factor has several components which all affect the commercialisation of biotechnology: availability of finance, taxation levels and allowances, health and safety regulations, science base, feed-stock prices, and patent protection.

Present national R&D funding still reflects the power of science lobbies of the past, rather than the economic potential expected from present R&D. The current balance between the physical and biological sciences is still in favour of the former, and a reversal of this balance is overdue as the problem cannot be solved by ever-increasing science budgets.

Dr. Coleman's conclusions focused on the fact that the effectiveness of national strategies has been difficult to assess because the consequences of a particular policy cannot easily be separated from events which might have occurred even without government intervention. This was true, for example, with regard to the creation of new biotechnology companies. Increased resources for biotechnology from one of the large chemical corporations could be more effective than many new small companies.

Each country has to have its own approach to biotechnoloy, based on its strengths and industrial structure. Consensus on a "best" approach would not be appropriate. Dr. Coleman highlighted the importance of international co-operation in research, which was an excellent way to efficiently use scarce resources. There was a role for international organisations in promoting this, and Dr. Coleman contrasted the respective policies of the OECD and of UNIDO in this context.

The speaker regretted that international co-operation in biotechnology was limited because countries were too defensive and concerned with short-term interests. He stressed the United Kingdom's willingness to work with other countries in developing the scientific infrastructure of biotechnology.

Dicussions following Dr. Coleman's presentation focused on several key issues: the question of regulations and the importance of public acceptance of biotechnology emerged as major themes of the Workshop. It was noted that public acceptance is a key factor not only in affecting the regulatory situation, but also in developing markets for new products and processes, markets being at least as important for the spread of biotechnology as the often quoted "climate".

In addition, the implications for developing countries as a result of applications of biotechnology, were stressed, and joint research projects to deal with problems of these countries (tropical diseases, water pollution, malnutrition) were suggested. Such projects could be part of joint international action. One participant, however, emphasized that biotechnology would increase the difficulties of developing countries, because one of the main thrusts of biotechnology will be to reduce the trade-dependence of OECD countries in various crops and also, because success in biotechnology requires a first-class competence in many R&D sectors.

2. Setting strategic priorities

Dr. Shin Aoyama, Deputy Director, Life Science Division, Science and Technology Agency, Japan, gave a keynote address on the Japanese system, and policy formulation in Japan. He outlined the spending on the life sciences in Japan and the Japanese view of the impact of biotechnology on industrial structures.

Dr. Aoyama described the government mechanisms currently in place to support the life sciences in Japan, including the research interests of the various agencies, as well as the way in which policy decisions are arrived at.

Biotechnology policies are formulated in the Council for Science and Technology which report to the Prime Minister. Ministerial meetings can direct precise policies. Presently, five key technologies are receiving priority support.

Questions and comments following Dr. Aoyama's presentation focused on Japan's Human Frontier Science Programme and how it would be able to encourage international co-operation. Delegates praised Japan's political commitment to basic research, and its recognition that future progress in biology must have a multinational basis. Some asked whether Japan intended to stimulate international programmes particularly in the five priority areas, and whether the presidency of the Human Frontier Programme would rotate between countries. Dr. Aoyama replied that the Programme was at an early stage and that details were still under consideration and being discussed at high levels by his government.

Several delegates contrasted their national experience in government support for biotechnology and strategic priority setting, with the experience of Japan and other countries.

A delegate from Italy pointed out that it was very difficult to change "climates" and that in his country, there was little hope for a spontaneous network of biotechnology companies to come into existence. Hence, governments should intervene strongly, set up R&D centres and tell scientists to focus on applied R&D. Another delegate (USA), however, warned that the transition from research to application was shorter in this than in other technologies (2-3 years), and that governments, rather than trying to direct this transition, should be generally supportive of industry and help to set the right climate. Representatives from France, Germany, Finland, and Sweden spoke of their country's policies which were somewhere between those two positions. One of the main objectives of government policies is to improve the interface between research and industry, at the earliest possible phase (Finland), and with regard particularly to traditional, and not only top-level industries (France).

3. The commercialisation of government and university research — issues in the public acceptance of biotechnology

Dr. Karl Heusler, Head of Central Function Research, Ciba-Geigy AG, Basle, addressed both topics. Regarding industry-university links, he asked why we were now speaking of problems, when biotechnology came from university laboratories like so many pharmaceutical inventions before that did not seem to have led to problems. The reason was that genetic engineering had opened up unlimited opportunities in biotechnology, provoking considerable excitement and a tendency to forget the difficulties associated with commercialisation. For co-operative projects to be successful, there must first be a common interest (to earn as much money as possible is not sufficient); second, a mutual benefit; third, mutual trust and recognition of capabilities; fourth, communication, and fifth, a mutual commitment (agreements from which either partner could withdraw, are doomed to fail). On many of these conditions, there have been conflicts between the aims of universities and industry.

The role of government policy should thus be to maximise the opportunities for co-operation, and to try to reduce the areas of conflict. Important components of such an approach are support for centres of excellence in the universities, and mechanisms to reduce the friction between the need for publication in academia and the need for secrecy in industry. Improved patent protection, and particularly the introduction of a grace period, are essential. This, as well as measures to encourage investment, should allow industry-university co-operation to develop. In any event, outside financial support is not the most critical issue as viable industry-university projects can do without such support.

Turning to public acceptance, Dr. Heusler stressed that modern biotechnology is extremely complex and that anything unknown is, naturally, frightening. In the past, Dr. Heusler believed that an intensive information compaign explaining the principles and achievements of biotechnology in simple and attractive terms would convince the public. Numerous campaigns have taken place, more and more people are informed, but the fear, instead of disappearing, has rather increased. In fact, over-information has created "cognitive stress" because of the large intellectual effort necessary to grasp the importance of this technology. As factual information is no longer in demand, it would be more effective to try to develop public confidence.

The challenge for governments is to develop public confidence in their risk management framework by demonstrating that technological development can be reliably controlled and, if necessary, constrained, to avoid over-information, to show that nothing is without risk and that benefits and risks are two sides of the same coin, and that projects where risks have become more important than expected benefits, have been abandoned.

Comments following Dr. Heusler's presentation focused first on industry-university links. Delegates from Yugoslavia and Japan presented a critical assessment of the changing, but still not completely satisfactory situation in their respective countries.

There was agreement that open and easy communication between industry and the university is essential in biotechnology, and that governments have a catalytic role to play in encouraging links, but also, that there should be no permanent government involvement in such links.

Government programmes should thus be designed to improve university centres of excellence — Germany's delegate mentioned his country's good experience in this respect — and to encourage links based on this excellence and on a commitment to common goals. When based on such conditions, co-operative agreements are more likely to survive in the long run than programmes based solely on funding for a specific project.

The issue of public acceptance of biotechnology dominated the comments. While many of the delegates accepted the concept of "over-information", they also suggested some specific areas where more public information is necessary, in particular, the fact that biotechnology allows for technical solutions that are safer and more predictable than "conventional" technologies.

There were several comments on the need to develop more credibility for risk management systems. It was suggested that for the public to have confidence in the management of the risks of biotechnology, the risk management agencies must not be connected to the "advocates" of the technology. Independence and impartiality have to be maintained. An analogy was made to the nuclear power industry, which suffered a loss of credibility because the risk management system was seen to be too closely related to the industry advocates.

4. Training and mobilisation of human resources — free flow and exchange of scientific information

Professor Daniel Thomas, Director of France's "Programme Mobilisateur-essor des Biotechnologies", spoke on training as well as the free flow of scientific information, demonstrating that both are interdependent. Professor Thomas emphasized that there is no way to train an ideal biotechnologist who would satisfy all possible requirements. One problem is the rapid progress in many underlying scientific disciplines, which means that close interaction between training and high calibre research is essential. Training has to be multidisciplinary. Amongst other disciplines, biology has to be integrated with engineering training which has been difficult in many traditional educational systems, including that of France. The fields where research is advancing particularly fast and hence where training methods have to advance as well, are: genetic engineering, microbiology, proteins, fermentation, bio-engineering, plant molecular biology, immunology. Speaking of the interactions between training and industrial manpower needs, Professor Thomas stressed that biotechnology training is needed not only for high-technology companies, but for traditional industries as well, and that the basic understanding of biotechnology in the labour force has to be improved at all levels.

Turning to the free flow of scientific information, Professor Thomas mentioned the danger that national programmes might divert too much free fundamental research towards short-term, technological objectives. He warned that there seem to be fewer scientific conferences where biotechnology researchers freely exchange scientific information; and more and more meetings on commercial or related biotechnology aspects. Moreover, an increase of preferential links between high quality university laboratories and industry seem to have led to a certain "privatisation" of scientific discoveries in biotechnology. However, free dissemination of scientific results through the scientific press, and competition between scientists are essential for the advancement of basic knowledge. Better patent protection, particularly a "grace period" which scientists are now asking for, would encourage this scientific exchange. The fastest and most efficient exchange of scientific and technological knowledge would be through personal mobility of highly skilled personnel — another way of linking the free flow of science to the training of biotechnologists.

In the *ensuing debate*, delegates underlined the importance of life science training in the education of all students, particularly those in faculties of engineering, and identified bioprocess-engineering as a priority area which was lacking in many university programmes.

On the subject of scientific information flow, one delegate asked whether scientists from the most advanced country, or countries, were necessarily interested in free information flow. Why, in such a competitive business, should they share information with the whole world? Other delegates underlined that the support of, and free access to data banks, and the free exchange of research equipment, were important components of the free flow of scientific information. It was appropriate for international bodies to address these issues.

5. Safety and regulations, R&D and international co-operation in biotechnology

In his presentation on this subject, Dr. David Kingsbury, Chairman of the Domestic Policy Council, Biotechnology Working Group, and Assistant Director at the US National Science Foundation, outlined the United States' experience and approach in dealing with the regulation and safety of biotechnology. He stressed the need for international co-operation on this issue in order to avoid the creation of trade barriers and underlined that regulatory decisions had to be scientifically based, which meant amongst others, that adequate R&D resources had to be devoted to the environmental sciences.

Part of the planning for this effort was a proposal to establish an international data base related to the introduction of new organisms into the environment. Initially, this has been a joint activity of the United States and the Commission of the European Communities, but Dr. Kingsbury offered to extend the planning in order to include at least all OECD Member countries. It has been envisaged to set up a data base of seven interlinked data files (taxonomy, literature, organisms, release events, a directory of related information sources, and an electronic bulletin board). This would include data from the last 100 years.

Turning from the information needs of the scientific community to that of the broader public, Dr. Kingsbury expressed the view that the term "biotechnology" should be clarified or not be used any longer. The word has become a millstone around the "neck of industry and governments", because the definition has been broadened gradually to include numerous techniques which were safely used for decades, without the specific attention they now receive and which could drag them into oversight and regulation problems. Biotechnology was not a unitary entity, but an enabling technology with broad and diverse applications. It was the use of a single, unprecise term to describe so many diverse activities, which had raised public concern. In addition, Dr. Kingsbury proposed efforts to provide more information on the accomplishments and benefits of the new products and processes, but cautioned that there should be realistic assessments of future applications. In conclusion, Dr. Kingsbury called for more research to establish a scientific base for rational regulation. As well, he cautioned against "over- or under-regulation" which can have consequences on research, commercialisation and public acceptance. He hoped that eventually it would be possible to eliminate the present case by case analysis of regulation, and develop an international agreement on classification and exemption from regulation for many of the new products.

Dr. Kingsbury's presentation sparked considerable debate over the word "biotechnology" and the scope of its definition. There was agreement that applying the word as a blanket over a number of unrelated and non-controversial technologies could be counterproductive.

However, several participants (France, EEC) refused to have the term abandoned completely; policy-makers had to learn to cope with its negative as well as its positive implications, and a very broad definition could also be useful as it could lead to the "banalisation" of biotechnology in the eyes of politicians and the public.

There was support for international efforts towards the development of appropriate and unambiguous regulations. There was general agreement that regulation must be product-based, but that there should also be mechanisms in place (many of which already exist) to ensure safe industrial practices. The subject for priority attention in this field was risk assessment and risk management. While risk management must, by its nature, be a national process, risk assessment criteria and methods were excellent areas for international action.

Risk assessment R&D, which presently does not attract many scientists, should be formulated to include basic research, which would make international co-operation possible.

After the keynote speeches and discussions, the Workshop decided to form three parallel Working Groups to study three sets of topics, with the aim of formulating a number of policy recommendations.

The three Groups had the following terms of reference:

— Public Information and Acceptance of Biotechnology (Chairman: Dr. John Cohrssen, Council of Environmental Quality, Washington, D.C., United States);
— Scientific Information and Data Bases — A Challenge for International Co-operation (Chairman: Dr. Brian Richards, British Biotechnology, Oxford, United Kingdom);

— Climate for Commercialisation of Biotechnology and Impacts on International Interactions (Chairman: Dr. Mark Cantley, Concertation Unit for Biotechnology in Europe, EEC, Brussels).

II. POLICY RECOMMENDATIONS

The following policy recommendations have been formulated by experts who spoke in a personal and informal capacity. Thus, they do not necessarily represent the opinion of the organisations, agencies and governments with which these experts are affiliated, nor do they commit the OECD or any of its Member countries.

1. The climate for commercialization: regulations and patents

The commercialization of biotechnology typically requires *long-term*, *high risk* investment, oriented to *international* markets.

For all companies and countries, key factors in the present climate for commercialization are the regulatory environment and protection for intellectual property.

Appropriate and unambiguous regulations are one of the most important factors currently influencing the climate for commercialization of biotechnology. Adequate patent protection is vital to the development and international flow of technology and to scientific co-operation. Absence of protection impedes international technology transfer by increasing uncertainties related to investment decisions. Patent protection benefits extend not only to the most advanced countries, but to developing as well, where patent legislation has been demonstrated as being effective in facilitating scientific and technological progress.

Recommendations

In order to reduce the uncertainties, it is recommended that Member countries:

a) Base regulations on rationally selected categories (typically focusing on products rather than processes) and on scientific criteria, and facilitate international harmonization of such criteria;

b) Make maximum use of international pooling of scientific information relevant to regulation, and identify needs for further research;

c) Build on OECD's previous work on patent protection in biotechnology and develop criteria for adequate protection of intellectual property in the context of biotechnology particularly for plants and animals, in consultation with appropriate international bodies ; and

d) Examine ways to facilitate patent protection by simplifying procedures and reducing often prohibitive cumulative costs of obtaining and maintaining worldwide protection (which pose particular problems for small firms).

2. Public acceptance of biotechnology

Public confidence in biotechnology products and processes will have a significant impact on the pace of commercialization. Questions of safety and ethics are obscuring the benefits, both economic and social, that will be derived from the applications of biotechnology. Medical research, diagnostics and treatment, as well as environmental protection and waste treatment may benefit from applications of biotechnology. Public acceptance is important and awareness programmes will be an important component of any strategy to encourage biotechnology development.

Recommendations

In order to maintain public confidence, it is recommended that Member countries:

a) Make sure that the public is aware that a regulatory framework, which can appropriately deal with issues of public concern is in place for biotechnology processes and products;

b) Make information available on benefits and possible risks associated with biotechnology and establish mechanisms for public participation in developing the process of identifying and assessing both;

c) Encourage agencies to set aside adequate resources for activities relating to public awareness; and

d) Exchange information on the benefits and on the scientific evaluation of possible risks in the interest of establishing common approaches.

3. The application of biotechnology to public objectives

Some applications of biotechnology may not be of great commercial interest, but may be of considerable public interest and value.

Recommendation

In order to encourage beneficial biotechnology applications which may not lie within the realm of commercial activities, it is recommended that Member countries:

a) Promote the application of biotechnology aimed at public objectives which may not presently be undertaken in the commercial arena, such as the development of "orphan" drugs, the reduction of animal use in testing and experiments, or the protection of the environment and pollution abatement.

4. Needs for smaller companies and of less-advanced countries: mecanisms to facilitate biotechnology development

The long-term, high risk, and international characteristics of biotechnology may create particular difficulties for smaller and medium-sized companies, and for countries less advanced in biotechnology. Contemporary biotechnology development includes important and distinct roles for both small and large enterprises. The large enterprises typically have the financial resources, international awareness, management capability and in-house infrastructure enabling them to cope with large-scale production, and to gain access to global markets.

Small companies typically provide specialized and dynamic research, but may be unable to satisfy their needs for information, particularly on market factors (opportunities, competition, etc.).

A similar problem pertains to many financial agencies and banks which, while international in their financial awareness and activities, do not have the technological sophistication to evaluate biotechnology proposals in-house.

In a number of OECD countries, various agencies have already fulfilled a useful role in addressing these problems.

Recommendations

In order to facilitate biotechnology development and commercialization, it is recommended that Member countries, when necessary:

a) Develop the role of various agencies, as providers of information, or as "gateways" to appropriate information sources elsewhere: with particular reference to the needs of small companies for market information, for international information (e.g. how to establish operations, find partners, meet local requirement), and for other information; and

b) Develop the role of such agencies, with particular reference to the needs of financial institutions for information relevant to their assessment of biotechnology projects.

5. Scientific information and data bases

Individual Member countries are compiling data bases in sciences related to biotechnology which can be of international utility. They will be valuable for scientific research and many other science based goals, particularly regulation purposes such as assessment of possible risks from the introduction into the environment of genetically modified organisms. To facilitate their international usefullness, consideration should be given to the means of access from electronic networks and the ability to include data from all OECD countries. This constitutes a unique challenge to international co-operation.

A means for recognizing scientists' contributions and for ensuring the quality of the input should be found. OECD may be able to encourage the development of a peer review system analogous to the systems used to review scientific publications.

Recommendations

In order to strengthen the international usefulness of biotechnology data bases, it is recommended that Member countries and the OECD consider:

a) Facilitating access, inputing and updating data bases on an international basis;

b) Ensuring the means for recognition and peer review of contribution to data bases;

c) Facilitating the international exchange of data for regulation purposes;

d) Addressing the cost problems of such data systems on an international basis; and

e) Providing, where appropriate, the assurance of confidentiality of proprietary information.

6. International trade and biotechnology

The outlook of the biotechnology industries is international. The prospect of free trade is a powerful stimulus to biotechnology development and commercialization, while trade restrictions could discourage such development and commercialization.

Recommendations

In order to facilitate biotechnology development and commercialization, it is recommended that Member countries:

a) Make every effort to avoid the use of regulations as non-tariff barriers to trade; or conversely, as an instrument of competition by adopting "easiest" rules; and

b) Develop and harmonize standards of relevance to commercialization and trade in biotechnology (e.g. for purity of materials, test protocols, etc., where these do not already exist).

III. KEYNOTE ADDRESSES

Opening address

The Hon. Frank Oberle, P.C., M.P.
Minister of State for Science and Technology

On behalf of Prime Minister Brian Mulroney, I am pleased to welcome our distinguished guests to Canada. It is a great pleasure for us to co-host this workshop with the Organisation for Economic Co-operation and Development.

I am sure I also speak on behalf of the participants from Canada who join with you in this important workshop. We hope your experience with us will be pleasant, rewarding, and productive. We in Canada share with you great interest in the promise which biotechnology holds as a major contributor to economic and social progress.

I would like to begin addressing this subject in the larger context. Biotechnology is undoubtedly among the world's oldest sciences — but is is still in its infancy. In a recent address on science and technology our Prime Minister said:

"Perhaps no area is more challenging
and profoundly dramatic than that
of biotechnology ..."

Indeed, biotechnology is growing out of its infancy and into the new technological age in a most spectacular fashion. We in Canada recognize as well that most countries in the world today face important challenges in trade and economic development.

All of us are turning to science and technology as the means to improve our competitive position in the world's marketplaces. For us in Canada, biotechnology shines as one of the most promising fields, one which offers great hope, and potential to add to our nation's economy.

For us, biotechnology provides new ways to produce higher value-added products from the natural resources which are the basis of much of our economic structure. As you all know, Canada — in comparison to its partners in the OECD universe — has an industrial base which is unique. It is shaped by the nature of our country — our territory is the second largest in the world.

We are the custodians of 20 per cent of the world's fresh water. We own and manage one of the world's richest storehouses of natural resources: minerals, forests, grasslands. We are surrounded by three oceans. All these resources provide new opportunities — but present, as well, many demanding challenges. The nature of our country has, in many ways, shaped the attitudes of our people and the value system of our society.

We, as you can understand, had to be mindful of these attitudes as we began to develop policies and commit to strategies in order to come to terms with the "technological revolution" which is showing us the dawning of a new age.

Of course, we can learn from others and contribute our own findings to the base of knowledge which is becoming the currency of the new age. But, like everyone else, we have our own strengths upon which to build, and our own difficulties to overcome.

I think it may be useful to provide you with a brief overview of what we are doing in Canada with science and technology. We have tried to make our solutions appropriate to our unique circumstances, and there may be some principles which could be adapted elsewhere in the world.

Starting two years ago, our government has tried to bring together all sectors to develop a first-time national science & technology policy. We organized a national forum involving industry, labour, universities, provincial governments, the financial sector, and last month I joined with my provincial counterparts to sign a national science and technology policy. This policy is designed to correct some historic problems such as the disparity of some of our remote regions and the lack of co-ordination of the efforts of the various players in the field of science and technology.

We have formed a council of science and technology ministers to demonstrate national unity in co-ordinated implementation of the policy, and our Prime Minister recently announced the establishment of a new national advisory board on S&T which he will personally chair and which is made up of some of the most eminent persons in our country.

Last year, the House of Commons created a standing committee on research, science and technology. We have moved science and technology from the periphery to the very centre of government decision-making.

To complement that policy and to guide federal participation in the implementation of the policy, a few weeks ago I announced on behalf of the Federal Government the broad outline of what we consider to be a practical agenda for Canada's science and technology strategy.

It emphasizes:

— Industrial innovation and the diffusion of technology;
— The development of strategic technologies;
— Effective management of the $4 billion spent annually by the Federal Government in science and technology;
— Development of highly-qualified human resources; and lastly,
— Measures to develop a Canadian culture which cherishes and rewards scientific and technological endeavour and excellence.

We have named this plan "InnovAction". Although it will be modest in comparison, for instance to the Human Science Frontier Program in Japan and the Eureka and Esprit programs in Europe, it is intended to capture the spirit of our citizens on whose support the success of our program depends.

In the development of the strategic technologies as part of the InnovAction program, we recognize biotechnology as the area toward which much of our effort will be focused. We are a country with a relatively small population base who can no longer afford on our own to support every project or every initiative in every discipline of science.

Like most of our partners in the OECD, we have learned that we cannot afford any longer to work in a policy vacuum, leaving everything to chance, hoping that the mix of choices a decade ago will still fit, or hoping that someone, somewhere, will make the right decisions. We also learned that we cannot afford to make decisions based solely on inappropriate international comparisons. Rather we are making decisions appropriate to *Canada's* strengths, decisions which support *our* values and which advance *our* priorities.

I am sure many of you representing other nations, appreciate and share these sentiments.

Above all, we have realized that we must act quickly to stay in control of the trends which will so profoundly affect the future of all mankind.

Before I go on, I owe it to you to explain a favourite Canadian sport — self-criticism. In speaking with Canadian colleagues, you probably have heard a long list of our troubles and wœs. We Canadians have a tendency to be self-deprecating searchers for greener pastures. True, while we all know we have a lot of work ahead, the facts would indicate that we Canadians have much to be proud of. We are among world leaders in many areas of technological progress.

This year, for instance, we celebrate a quarter century of Canada in space. We were the third country to establish a space program. We were first in fact to develop and deploy a commercial communication satellite solely for our own use. We have built a highly competitive aerospace industry, a nuclear industry, a transportation sector.

Now it is true that in accordance with the normal criterion in use to compare R&D spending performance — that is on a percentage of gross domestic product — Canada shows up on a lower end of the scale. But perhaps and again considering our own peculiar circumstance it would be fair to make some comparisons of spending on a *per capita* basis.

Here we discover that in regard to expenditure on R&D for the advancement of knowledge only Germany and New Zealand have *greater per capita funding* than Canada — in this vital area, Canada is ahead of the US, ahead of Japan, ahead of France and Great Britain. In regard to research and development for health and welfare, only Germany and Sweden spend more on a *per capita* basis than does Canada.

In regard to research and development for economic development, only France, Germany and New Zealand spend more. Even in defence and space R&D, Canada ranks in the upper third, seventh on the list, Canadians can say with pride that at least in some comparisons we stand at the top of the list among the world leaders in these important areas which fit Canada's international image as a humanitarian, educated, compassionate enlightened nation. I hope that your own regard for Canada reflects the image and reality of the kind of country which we wish Canada to be among the community of nations.

Now in all of this, I have been particularly interested in the way in which biotechnology is emerging with its new and varied manifestations. You know, I wonder sometimes what the world would be like if over time we would have directed more of our ingenuity to developing biotechnology, as a means to do the things for which we have, instead, created products and solutions which are hostile to nature, and our environment.

The significant new developments in biotechnology, added to the classic techniques which have been used for thousands of years, will give us new opportunities for responsible stewardship of our environment, and natural resources. Certainly some care is necessary, and I hope that conferences like this will address the critical questions of ethics and regulations. Canada is concerned about these issues and we seek to co-operate with others in establishing necessary codes of conduct that respect shared human values and dignity.

Despite the potential which today seems to be so obvious, biotechnology has been very much a neglected field in Canada until quite recently. At the beginning of the 1980s, there were only a handful of firms engaged in biotechnology. Now in 1987, almost one hundred firms are developing and specializing in biotechnology.

In 1981, a private sector task force on biotechnology chaired by Maurice Brossard reported that Canada lacked scientific and industrial strength in that field. Further, the task force concluded that biotechnology, including genetic engineering, enzymes, fused-cell techniques, plant-cell culture and process and systems engineering would have major impacts on a number of Canadian industries. In particular, it underlined the importance of biotechnology to innovation in Canada's resource industries. The task force urged the establishment of a national long-term strategy to building a biotechnology capacity through a multi-sectoral program with clear federal leadership.

Now, just six years later, joint action by many Canadian sectors has built a scientific infrastructure in this technological area that provides us with the foundation upon which we can work and build. At the present time, there are networks of university, industry and government — involving 1 500 researchers and managers — assisted by a cost sharing program operated by our National Research Council.

We have seen that many small, technology-intensive firms have started up. These firms carry out research and testing of biotechnological processes and assist in the integration of innovations in a wide range of industrial operations. These one hundred private sector firms are currently spending about $90 million annually on research and development. An even greater number of industrial firms which process resources or manufacture products maintain their own in-house biotechnology research facilities.

I can assure you of the Canadian government's continued support toward a national strategic approach to realizing fully biotechnology's promise. I would like to mention a few examples of the impressive results which the private sector has produced in the last few years:

— Rhizotec incorporated of Quebec has used a microbial fertilizer technology developed at Laval University to produce millions of trees which grow rapidly to reforest poor soils.
— Denison Mines Limited of Elliot lake in Ontario has established a test facility for underground in-place biological leaching of uranium ore; in British Columbia, Giant Bay Resources is bioleaching gold and silver ores.
— Nova Biotechnology Limited in Nova Scotia has evolved from a nursery operation growing blueberries into a tissue-culture facility using state-of-the-art technology.

In 1985-86, the Natural Sciences and Engineering Research Council awarded $17 million in research and training grants for biotechnology to researchers in our universities. Several universities have launched programs and our granting councils support professorships in biotechnology. The Medical Research Council (MRC) is spending $15 million for biotechnology-related research, and is sponsoring training programs. From 1975 to 1983, Canadian universities quadrupled their annual output of graduate-level professionals in biotechnological disciplines.

The flagship of our National Research Council's effort is the Biotechnology Research Institute in Montreal which will be officially opened soon (May 1987). It is a world-class facility, with an operating budget of over $30 million and a permanent staff of over two hundred persons, linked with researchers across the country — indeed, around the world. It is open to domestic and international joint ventures. The Montreal facility joins strong NRC teams in Saskatoon at the Plant Biotechnology Institute and the Division of Biological Sciences in Ottawa.

Dr. John Evans and the members of my National Biotechnology Adivsory Committee have noted that federal initiatives have significantly strengthened the research infrastructure for biotechnology, employment and economic development. They are particularly important in retaining our best scientists and attracting talented people to Canada. If you were to ask me do I think Canada is doing enough R&D in biotechnology, I would emphatically reply "no".

We cannot assess progress simply by looking at where we are now and where we were in 1981. In spite of the respectable increases in Canadian biotechnology investments, we have to assess our progress relative to other countries, and changing circumstances.

Understanding that biotechnology is not only multidisciplinary, but also that its knowledge base resides in many parts of the globe, we are looking for opportunities to form strategic alliances with other countries particularly with our friends who join us in membership in the OECD.

We in Canada recognize that we can simultaneously be both a competitor and a partner in co-operative efforts. We recognize, as I am certain you do, that it is essential that every country share in the prosperity and progress as we move into the 21st century as co-tenants of this planet.

Future observers will be passing judgment on us all, as does each generation on those who have gone before them. We in Canada are attempting to employ our energies and political talents so that in passing judgment, the future will say about us: "They made the right decisions. They had the right priorities. They did the right things".

You join me, I know, with a similar dedication. In closing, I would like to address my thanks to you for inviting me to this important conference. It is yet another indication of the spirit of co-operation shown by Members of the OECD.

National policies and programmes in biotechnology

Dr. R.F. Coleman, Government Chemist,
Department of Trade and Industry, United Kingdom

The rapidly increasing economic impact of biotechnology, as well as the safety and ethical issues raised by our ability to manipulate the basic life processes, has ensured that most national and international bodies develop views on the subject and this usually leads to a defined policy. The OECD is no exception. The excellent publications produced after careful deliberation by experts on, first of all, trends, secondly patents and, thirdly, safety, have been a major influence on national and company policies. The background paper we have received for this Workshop. "Biotechnology R&D — National Policy Issues and Responses" by Nancy Field, is another key document which will be widely used as a source of material in the future.

The purpose of my presentation is to take the information available in that document (supplemented in some specific areas) as a basis for the comparison of national responses to this important emerging technology called biotechnology. The purpose of our discussion is not to achieve unanimity of approach by government in all areas. Clearly, that would be impossible because important though biotechnology may be to the future, economic performance in a single sector is unlikely to cause a nation to change its political philosophy or overall industrial policy. Nevertheless, there are issues which are common to all and we can benefit from a closer consideration of the national responses to biotechnology as these will surely provide important signals for the future.

In my presentation, I will compare the national responses of several OECD countries. It is not as comprehensive as the OECD background paper, as it does not draw on information from all OECD countries. However, I have tried to include some comment from most of the major players in the game and to include examples of countries with different attitudes to industrial development. I am grateful to the colleagues in other countries who have provided information which I have included in this paper. I hope I have used the information correctly and I apologise for not referring to all the countries present at this Workshop. That would have been impossible in a short paper with only thirty minutes for presentation which aimed to provoke discussion rather than merely catalogue the past.

I will consider national responses in biotechnology under four separate headings. First, the economic and social climate influencing attitudes to industry and high-technology industries in particular. Second, the public support for the basic sciences underpinning biotechnology and the exploitation of developments arising from the science base. Third, policies relating to the supply of trained manpower and, fourth, the identification of national industrial targets in biotechnology. These are not the only areas in which some governments have developed specific policies but other speakers will be picking up other issues later in the workshop.

I. CLIMATE

May I remind you that the majority of the ten factors affecting the commercialization of biotechnology, identified by the US Office of Technology Assessment Report in 1984, were those generally pertaining to the development of high-technology industries and not specific to biotechnology. I would agree that the general climate in which industry operates in a particular country has a greater influence than any specific measures of support to an individual sector or technology. Thus, in general terms, the performance of a country in the commercialization of biotechnology is likely to follow the same pattern of performance for industry generally and it would be very difficult to bring about a major change in one sector, if the performance overall is weak. The important factors influencing the overall climate are:

— Availability of Finance;
— Corporation Tax and Allowances;
— Regulatory Regime;
— Science Base;
— Feedstock Prices; and
— Patent Law.

Finance

The availability of finance for companies is a major factor influencing the growth of the private sector. It is not only the absolute level of capital available to the companies but also the diversity of sources which can be important. I have given a personal opinion of the relative importance of three aggregate funding sources, namely, venture capital, internal company financing, including secured bank loans and government grants in six OECD countries in Table 1.

Table 1

RELATIVE IMPORTANCE OF FINANCE SOURCES
FOR BIOTECHNOLOGY[1]

	Venture capital	Retained profits	Government grants
United States	1	2	—
United Kingdom	2	1	3
Germany	—	1	2
Japan	—	1	2
Canada	3	2	1
Netherlands	3	1	2

1. Finance sources are ranked 1, 2, 3 in order of importance for each country.

There has been no lack of capital for worthwhile commercial ventures in the United States. The birth and growth of over 200 new biotechnology companies over the last 10 years, is a clear indication of a well-developed venture capital industry able to mobilise funds from private individuals and other organisations for those that can present realistic proposals with well-developed business plans. Equally important, there are many major chemical and health-care companies readily able to generate funding for diversification into new market areas requiring biotechnological techniques for the development of new products and processes to satisfy those market needs. In Europe, the UK most closely matches the US position, although the level of funding available for biotechnology is smaller, partially because of the lower expectations of short-term profitability likely from the application of biotechnological techniques. The major companies are able to raise the capital required and the British Venture Capital Association has about £250 million per annum available for financing new ventures in the UK and overseas. Approximately 50 new biotechnology companies have been formed in the UK and there is certainly adequate venture capital available for worthwhile business propositions. In the Federal Republic of Germany, the scene is dominated by the major chemical and health-care companies with little evidence of venture capital and, so far, very few new companies specialise in biotechnology.

In France, the situation is somewhat similar to that in Germany. However, a small number of specialist companies were set up with joint public and private fundings in the early days of the Biotechnology Mobilisation Programme. In Japan, venture capital is virtually unknown, with financing mainly provided by the banks. Few, if any, small companies have been formed but many established companies have diversified into the application of biotechnological techniques and established substantial R&D programmes, particularly aimed at the health-care sector.

Taxation

Taxation levels and allowances have a major influence on the availability of capital for new ventures. More than $1 billion was raised in the US through R&D limited partnerships in 1983 and 1984 and this can be largely attributed to the tax concessions. Partners can deduct as much as 85 per cent of their initial investment, because of losses in early years, to offset income from other sources which might be taxable at very high rates. The change in the taxation regime in the US, with respect to R&D partnerships, has now caused this source of funds to virtually dry up. In the UK, R&D expenditure is written off 100 per cent in the year in which it is incurred, so it therefore reduces the liability of companies to Corporation Tax. This is more generous than in most countries but in some (e.g. Australia), 150 per cent of additional R&D expenditure can be written off in the year, thereby further reducing the tax liability of a company. The Corporation Tax on profits varies considerably and can be a factor in establishing production facilities.

Health and safety regulations

The regulatory regime, covering products and processes, using biological techniques has a marked impact on the commercialization of developments from the biosciences. In general, companies like stability and certainty as far as possible and therefore a country which has a clearly enunciated set of regulations is almost certain to be preferred to those which have yet to develop their own regulations. Those responsible for framing legislation, should try to achieve a balance which does not inhibit innovation unnecessarily but adequately protects the consumer and worker from significant risks. As far as the UK is concerned, most observers accept that this balance has been achieved and a clear 'Plain Man's Guide" was issued in 1986 which has been widely acclaimed. The US regulations will be discussed in more depth later by

Dr. Kingsbury. The change introduced in 1986, which allowed US companies to export pharmaceutical and biological products, which had not yet received FDA clearance to certain specified countries, will be of considerable benefit to firms in the United States, and possibly delay the setting up of production facilities outside the US until fully justified by local markets.

The European Community will soon be proposing legislation to cover all Member countries. The known attitude of the Federal Republic of Germany and Denmark to the planned release of genetically engineered materials, will almost certainly make unanimity amongst Community countries difficult to achieve.

Science base

The rate at which discoveries, arising from basic research in molecular biology have moved through the applied research and development phases into new products and processes, is quite remarkable. Clearly, those countries with a strong base in biosciences, have been able to capitalise on these discoveries more readily than others. It is therefore not surprising that the United States, with its large expenditure in basic sciences, has made the most progress. The United Kingdom has also benefited from high quality basic research in this area, although the current rigidity in the funding system makes it difficult to switch funds into exploitable areas as rapidly as is necessary.

By contrast, Japan has found it necessary to licence-in technology because it has not historically deployed resources in basic life sciences. The current proposal from Japan for an international programme, called the Human Science Frontier Programme, aims to increase the resources devoted to basic biosciences. It is the Japanese view that without this increase in basic research, the breadth and depth of the science base throughout the world, will be inadequate for the development of future business activities. A somewhat similar view is put forward by the new biotechnology companies in the United States who argue that unless the US increases its funding in basic research, already large compared with other countries, the country is unlikely to continue to lead in the commercial exploitation of biotechnology.

Feedstock prices

The cost of raw materials can be a significant part of the cost of the end products. For most current products dependent on biotechnological techniques the raw material costs are not a major factor. However, as the techniques are applied to the protection of lower value products required in larger quantities, for example, biopolymers or sweeteners, the feedstock prices become a factor in locating production plants. In European Community countries, the Common Agricultural Policy controls the price of basic agricultural products and recently the price of sugar in the EEC has been approximately twice the "world price". In 1986, the pricing structure was modified so that sugar and starch for non-food use could be obtained at a special price. Without this change in pricing, it was becoming increasingly clear that production plants would be transferred to neighbouring countries with access to materials at lower prices and duty free access to the markets of Community countries.

The protection of the EC sugar beet industry from the cheaper high fructose syrup through a quota system, has also affected the development of large-scale production processes based on immobilised enzymes. The early lead in Europe has now been overtaken by developments in the United States because of the much larger markets upon to producers.

Patent law

We should not forget the expert report on Patents prepared by OECD, "Biotechnology and Patent Protection — An Individual Review" by Beier, Crespi and Straus, which drew attention to the variations in Patent Law in OECD countries. Two key issues are the grace period allowed in some countries, such as the US, between invention and patent filing and patent protection of plant varieties. The academic community world-wide appears to be in favour of a grace period. However, it adds complications to establishing priority dates for discoveries and the industrial sector is not unanimous on the need for change. In Europe, plant varieties cannot be patented, unlike the position in the US. It is now generally accepted in the UK at least that patent protection must be extended to plants, as the Plant Varieties Act does not give adequate protection to those introducing major changes to plants through gene transfer. Without such protection, large R&D expectations cannot be sustained.

II. R&D FUNDING

Basic research

The resources devoted to basic research generally do not reflect the perceived economic potential likely to arise in the longer term. The distribution of funds in the past has largely been dictated by pressure groups in the scientific community in the pursuit of new knowledge at the frontiers of science, for its own sake. The linking of scientific research in universities with education and training, which implies that "good" universities must be well funded for research, also distorts the distribution. It is difficult to change the pattern, as unexpected discoveries in a new area of science arise. I would suggest that a visitor from outer space would find the current balance between the physical and biological sciences surprising. With the most exciting areas of science now arising in the biosciences, and the major economic and social activities dependent upon advances in biology, a switch from physical sciences is overdue. The problem is unlikely to be solved, as in the past, by ever increasing science budgets to resource the exciting new areas whilst maintaining the existing levels of funding for established areas. All OECD countries with basic research programmes need to address this issue and accept the need for choices in basic science, as resources are unlikely to be found for all worthy proposals from the science community. The response of the Summit Countries to the Japanese proposal for an international collaboration programme called "Human Science Frontier" will be interesting and illuminating.

Strategic and applied research

The picture in applied research aimed at the commercialization of biotechnology is much clearer. With the exception of the US, most countries believe there is a need for Governments to provide support for biotechnology. In the UK, whilst expecting market forces to steer applied research, it is accepted that the market is partially distorted and public support for longer-term, risky projects, is appropriate. Perhaps for different reasons, many other countries have come to a similar view that limited support is appropriate from public sources but the majority of applied research should be from private sources.

a) Federal Republic of Germany

The FRG programme of applied biology and biotechnology, is directed towards the following goals:

— Scientific and technical excellence is to be vigorously encouraged and facilitated;
— The prerequisites for innovation in industry are to be improved;
— Research and development projects in specific fields of Government provision for the future are to be promoted;
— Opportunities and risks are to be assessed, existing safety regulations are to be further developed, and new safety aspects and ethical issues are to be tackled at an early stage; and
— Promotion is to be given to the younger generation of scientists in the field of biotechnology.

This has led to a range of support measures. The Society for Biotechnological Research (GBF) was established at Braunschweig as a national research centre, able to tackle a wide range of interdisciplinary projects. Other centres are funded from national and regional funds.

Three centres for genetic engineering have been established in Cologne. Heidelberg and Munich jointly with companies, but with majority funding from the Government. A programme aimed at helping companies use biotechnological processes, and to stimulate new companies, has been introduced. Grants of up to 40 per cent of approved costs (up to a maximum of DM 600 000 per company) are available for product and process development. To encourage collaboration between industry and research centres, up to 50 per cent grants are available for pre-competition studies in specific areas, e.g. cell culture and cell fusion techniques, enzyme biotechnology and bioprocess engineering.

Expenditure in Federal Republic of Germany is summarised in Table 2.

Table 2

PUBLIC BIOTECHNOLOGY R&D EXPENDITURE IN THE FRG — 1986

Million dollars

Bundesministerium für Forschung und Technologie (BMFT)	
Basic Funding for National Centres	33
Genetic Engineering Centres and Priority Projects	1
Indirect Measures	10
Co-operative Research	41
Total	97

b) Japan

The Ministry of International Trade and Industry (MITI) has been promoting biotechnology in Japan since 1981; in addition the Agency of Science and Technology (STA),

88

the Ministry of Agriculture, Forestry and Fisheries (MAFF) and the Ministry of Health and Welfare have been enthusiastic promoters of biotechnology in the relevant areas.

In conjunction with the Bio-industry Development Centre (BIDEC), MITI has actively supported improvements in the infrastructure for biotechnology, including providing a focus for the private sector, establishing gene banks and databases and encouraging standardization in the equipment required by industry.

The R&D support from MITI is partially in a 10 year programme on basic technologies for future industries, namely bio-reactor development; large-scale cell culture and the utilisation of R-DNA techniques. Assistance is also available to develop established industries in the production of fine chemicals, amino acids and water treatment. The public support for R&D in Japan is indicated in Table 3.

Table 3

PUBLIC BIOTECHNOLOGY R&D EXPENDITURE IN JAPAN — 1986

Million dollars

MITI	
Promotion of Technological Development	27
Underpinning Technological Development	5
Other	2
MAFF	
Breeding for 21st Century	12
Food Industry	4
STA	
Life Sciences	80
Total	130

c) *Canada*

The biotechnology industry in Canada is relatively young but growing and being actively promoted by Federal and Provincial agencies. The expenditure from public funds in support of biotechnology is indicated in Table 4 and is the dominant source of funds in this area in Canada.

The National Biotechnology Advisory Committee was established in 1983 to advise the Minister of State for Science and Technology and to develop an overall national strategy. A network of NRC laboratories has been established; Ottawa as a centre for advanced research; in Saskatoon for plant biotechnology and in Montreal for industrial R&D in collaboration with companies.

Over one third of the Canadian federal expenditure is in support of NRC activities — those mentioned above and industrial research grants. The other major funding area is agriculture.

Table 4

PUBLIC BIOTECHNOLOGY R&D EXPENDITURE IN CANADA — 1985-86

Million dollars

Department	
Agriculture	8.7
Forestry	0.4
Environment	1.0
Energy, Mines and Resources	1.0
National Research Council (NRC)	12.3
Natural Sciences and Engineering Research Council	4.3
Medical Research Council	0.8
Industrial Research Grants	7.5
Others	0.9
Provincial	13.1
Total	50.0

d) United Kingdom

The need for a cohesive R&D strategy in biotechnology was articulated in the recommendations of the "Spinks Report" (1980). This Report was compiled by a Working Party for Biotechnology, under the auspices of the Advisory Councils and the Royal Society, with a mandate to review prospects and to recommend action to facilitate British industrial development in biotechnology. The Spinks Report recommends that the "activities of Government Departments in relation to biotechnology, should be co-ordinated: a coherent programme of industrial research and development, involving industry, Government Research Establishments, universities and Research Councils, should be pursued ...". This objective was achieved by designating the Department of Trade and Industry as the lead Department and forming an Interdepartmental Committee on Biotechnology (ICBT) to co-ordinate where necessary the activities sponsored by individual Departments and provide a focus within Government for discussion and action. The ICBT has no funds of its own but persuades individual Departments to sponsor the agreed activities. The ICBT also provides a forum for discussion on international projects in biotechnology, including co-ordinating the UK response to proposals from the European Commission.

The science base, underpinning biotechnology, is the responsibility of the Research Councils. The Medical Research Council, the Agricultural and Food Research Council, the Science and Engineering Research Council and the Natural Environment Research Council all have substantial programmes of strategic research in biotechnology. The SERC has formed a Biotechnology Directorate which establishes priorities for research and actively solicits proposals in those areas.

In November 1982, the Secretary of State for the Department of Trade and Industry launched a campaign to foster the development of biotechnology in industry with the following objectives:

1. To raise the awareness of opportunities arising from biotechnology through publicity, provision of advice and selective grants to companies for feasibility studies by external consultants.

90

2. To raise the level of research and development in industry by the selective award of grants under the Support for Innovation Scheme.
3. To improve the infrastructure of biotechnology, for example through supporting industrially oriented work in culture collections and the development of information services for industry.
4. To develop a healthy climate in which biotechnology can flourish and remove obstacles and weaknesses in our system.
5. To identify and promote sectors of biotechnology that represent particularly important opportunities for the UK.

In order to implement these objectives, the Department set up a Biotechnology Unit, staffed by secondees from industry and civil servants. This mix of backgrounds amongst the staff of the Unit, has been particularly effective in ensuring that proposals for support from companies are properly appraised from the technical and commercial viewpoint.

The Ministry of Agriculture, Fisheries and Food, the Department of the Environment, the Department of Education and Science, and the University Grants Committee also support research and development in biotechnology in relation to their own specific policy objectives. The UK Government support for biotechnology is indicated in Table 5.

Table 5

PUBLIC BIOTECHNOLOGY R&D EXPENDITURE IN THE UK — 1986

Million dollars

Medical Research Council	43
Agricultural and Food Research Council	33
Natural Environment Research Council	2
Scientific and Engineering Research Council	6
University Grants Committee	5
Department of Trade and Industry	10
Ministry of Agriculture, Fisheries and Food	6
Other Government Departments	3
Total	108

e)   The Netherlands

In the early eighties, the Dutch Government implemented its policy for biotechnology in the form of two schemes:

— The innovation oriented stimulation programme for biotechnology (IOP-b) that was set up to help develop the R&D infrastructure; and
— The attention policy for biotechnology to support high-risk industrial R&D projects.

The IOPb has limited funds, about $20 million per year, which it uses to steer and stimulate the already-existing and regular financing schemes for university and institute research towards biotechnological goals, determined by discussion with industry.

The Ministry of Economic Affairs, which has overall responsibility for IOPb, provides additional funds to help industries direct their national research efforts to new areas of biotechnology to develop new products and processes. Grants up to a maximum of

45 per cent of the cost of new basic and application-oriented research programmes, are available.

f) Other countries

Many other countries provide support for biotechnology from public funds. *France* has a Mobilisation Programme co-ordinating the efforts of public research institutes, universities and industry aiming to substantially increase their share of world markets based on products derived from biotechnological processes.

The United States supports basic research in biosciences to a greater extent than all OECD countries taken together. This has been very beneficial in providing a basis for the new biotechnology companies, both in terms of exploitable research and high quality researchers able to move to product and process development.

Very approximate estimates of expenditure in biotechnology for the US, France and the Netherlands are indicated in Table 6.

Table 6

PUBLIC BIOTECHNOLOGY R&D EXPENDITURE IN OTHER
COUNTRIES — 1986

Million dollars

United States	Basic Research (NIH, NSF)	600
France	Mobilisation Programme	100
Netherlands	Innovation Strategy for Biotechnology	30

III. MANPOWER FOR BIOTECHNOLOGY

It is inevitable that the education system in all countries will lag behind industrial requirements if there is a spectacular change to manpower requirements. In the US and Europe, whilst there have been some shortages in limited areas, the availability of manpower of adequate quality and experience has not, so far, seriously hampered development. However, if biotechnology does take off at the predicted rate, shortages in the future may occur.

It is appropriate for international agencies to address manpower requirements and training and encourage national organisations to co-operate and help each other overcome deficiencies. The training programme supported by the European Commission is probably the most successful outcome of their biotechnology programmes to date.

As Professor Thomas will discuss training and human resources in greater depth later, I will not elaborate further at this stage.

IV. TARGETING POLICIES

In the UK, whilst we accept that government has a role to play in industrial development, it is primarily in developing the climate and infrastructure in which industry can flourish. Our experience indicates that the market will be a better guide to product and process developments. Nevertheless, the Department of Trade and Industry and the Science and Engineering Research Council have identified broad areas for strategic research which should be encouraged and deserve priority in the competition for limited funding. Additionally, collaborative research involving both industry and public sector research institutes has been particularly encouraged through larger grants to companies sharing in pre-competitive research projects.

It is my impression that Canada, France, Federal Republic of Germany and the Netherlands have a slightly more interventionist view and a greater government involvement in industrial strategies.

In Japan, MITI has a positive policy towards the bio-industry sector. This policy is agreed with industrial representatives in the Bio-industry Advisory Committee and implemented through the Bio-industry Office in MITI and the Bio-industry Development Centre (BIDEC).

Appropriately, the next presentation by Mr. Aoyama from Japan will give more information on strategy and further opportunity for discussion.

V. CONCLUSION

The effectiveness of particular national strategies is not easy to assess because it is difficult to separate the consequences of a particular policy from events which would have occurred in the absence of Government intervention.

Table 7 shows the number of biotechnology companies in five OECD countries. It indicates a growth in the number of specialist biotechnology companies in the US and, to a lesser extent, in the UK, and the diversification of many Japanese companies into biotechnology. By contrast, Germany and France have generated very few new companies but this is not necessarily a measure of commitment or potential success. Increased resources from large companies, such as Hœchst or Elf-Aquitaine, could be more effective than a dozen new small companies.

Table 7

BIOTECHNOLOGY COMPANIES AND SECTOR OF INTEREST

Sector	US	Japan	France	UK	FRG
Agriculture	73	12	5	15	2
Chemical	37	31	1	4	4
Diagnostics	141	15	3	10	6
Food	18	17	2	12	1
Pharmaceuticals	65	28	2	9	4
Veterinary	54	2	3	6	0
Total	388	105	16	56	17

In a recent issue of Biotechnology, Nigel Webb compared the incentives of three European countries for inward investment. Many countries have been trying to build up their industrial base by such methods as well as building on national companies. The incentives are indicated in Table 8.

Table 8

GOVERNMENT INCENTIVES IN THREE EUROPEAN COUNTRIES

Per cent

	Germany		The Netherlands		United Kingdom	
	Grants	Loans	Grants	Loans	Grants	Loans
Facilities						
Building	15	80	35	B	30	100
Equipment	25	B	35	B	30	B
R&D Assets	40	B	35	B	30	B
Expenses						
Production	14	0	0	0	0	0
R&D	20	0	45	60	25	0
Training	80	0	60	0	80	0

1. B denotes private bank loans are available. Many of these incentives are discretionary and only apply to certain zones. The percentages are maxima for qualified projects and are usually subject to limits.
Source: Nigel Webb, Biotechnology 1987.

The financial incentives have a marginal effect on the location of a subsidiary R&D or production centre, but other factors are very important. The overall climate for industry and access to skilled manpower are probably the dominant factors. On the basis of evidence available to date, multi-nationals and US biotechnology companies appear to be favouring the Netherlands and the UK as a location for expansion in Europe.

I have tried to present information to guide discussion at this workshop. Whilst we should not expect that the policies and priorities to be necessarily transferable across national boundaries, we can learn from each other. Additionally, with a better understanding of each other's policies and objectives, it may be possible to encourage more active collaboration and make better use of scarce resources. To date, international collaboration in biotechnology has been limited, possibly because we have been too defensive and more concerned with short-term advantages, rather than the longer-term opportunities. The UK is very willing to collaborate with other countries in developing the scientific infrastructure we all need and in pre-competitive research programmes.

Setting strategic priorities — the Japanese system

Shin Aoyama, Deputy Director, Life Sciences Division,
Research Co-ordination Bureau, Science and Technology Agency, Japan

First, I would like to mention that I often use the term "life sciences" instead of biotechnology in my presentation. The term "life sciences" is used traditionally throughout the discussions of the Government in Japan, and it is almost impossible to classify the Japanese R&D activities with the word biotechnology.

Here, "life sciences" covers science and technology for elucidating the mechanisms of life phenomena and various functions of living organisms, and for applying research results in medical treatment, agriculture, industry, environmental preservation, energy development, and to relevant aspects of human life. To my understanding, it is a somewhat broader concept than biotechnology.

I. LIFE SCIENCES R&D IN JAPAN

One of the most remarkable aspect of Japanese life sciences R&D is that the private sector shoulders 68.5 per cent of the R&D expenditure. Although the Government's role is strictly limited to lead the nation in work which cannot be shared by the private sector itself, the percentage of government and local government expenditure in this field is very much higher than that for science and technology as a whole (20 per cent). This fact may indicate that there still is a certain role for the Government to play in this field. The second aspect is that the total R&D expenditure is growing at a good pace, in parallel with that for the whole field.

Dividing the expenditure by research purpose, health science, medical treatment and products shares 70 per cent. Elucidation of life phenomena and biological functions share 8.8 per cent, and foodstuffs 6.5 per cent. In addition to this categorisation by purpose, recombinant DNA research, which may be spread over all purposes, shares 4.1 per cent of the life sciences expenditure, and it amounts to Y 36.2 billion with 2 300 researchers. Expenditure on recombinant DNA research has been rising at a double digit pace.

Figure 1.   LIFE SCIENCES R&D IN JAPAN

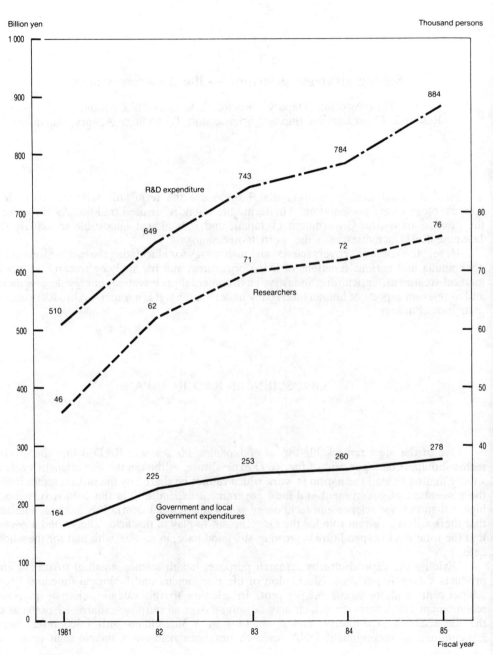

The private sector supported 68.5 per cent of R&D expenditure in 1985. The Government's role is to foster and enhance R&D which cannot be supported by the private sector.
Source: Bio-industry Development Centre.

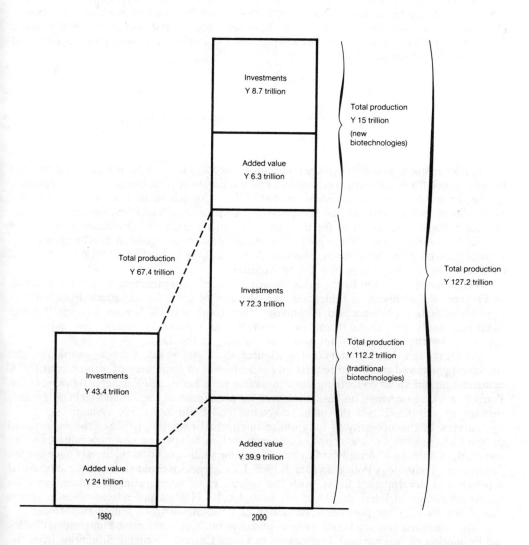

This figure shows the size of bio-industrial production.

Source: Bio-industry Development Centre.

II. IMPACT OF BIOTECHNOLOGY ON INDUSTRIAL STRUCTURE IN THE YEAR 2000

The Japanese biotechnology-related industry will grow to have a total production of Y 127 trillion in the year 2000. The total added-value of Y 46 trillion will be 10 per cent of total economic production, according to the Biotechnology Development Centre, a foundation of the bio-industry. The contribution of new technology such as recombinant DNA, cell fusion, and bio-reactors will be, however, about 15 per cent. This is based on the fact that Japan has a certain established traditional industries of fermentation, agricultural chemicals and pharmaceuticals.

III. FORMULATION OF GOVERNMENT POLICY

Fundamental and comprehensive policy on science and technology in Japan is formulated by the Council for Science and Technology, which is the supreme deliberative organ reporting to the Prime Minister, established in 1959. The Council is also responsible for the establishment of long-term and comprehensive goals for research and developemnt in science and technology, and measures for the promotion of research and development of special priority in order to achieve these goals. The Prime Minister is requested to fully respect any reports presented by the Council, and the Council is able to present additional views and comments on the inquiry by the Prime Minister.

The Council is chaired by the Prime Minister himself, and members include the Minister of Finance, the Minister of Education, the Minister of State for Economic Planning, the Minister of State for Science and Technology, the President of the Science Council of Japan (who represents the scholars) and five others of outstanding knowledge and experience appointed by the Prime Minister with the consent of the Diet.

As for the individual matters on life sciences, the Panel on Life Sciences was set up, and has investigated and submitted reports on fundamentals of promotion of recombinant DNA research, and on views concerning basic measures for a better aging society. At present the Panel is working on views on basic measures for promotion of science and technology and, through its sub-panels, on the immuno-system, and brain and nerve system.

In view of the necessity to fully utilize intellectual resources to make the society and people's life wealthier in the coming century as well as to open up new possibilities for the future, the Cabinet authorised the Council's recommendation entitled "General Guideline for Science and Technology Policy" in March 1986. Life sciences is listed as one of the substantial research and development fields, with the objectives of investigating life phenomena by taking advantage of latest developments in molecular biology and related areas in recent years, and looking for possible applications of new knowledge resulting therefrom.

The Guideline is based on the recommendation on "Comprehensive Fundamental Policy for Promotion of Science and Technology to Focus Current Changing Situations from the Long Term View". In the recommendation, life sciences are recognized as a leading and fundamental science and technology. It stresses the promotion of biological series of science and technology including elucidation of life phenomena at various levels of biology, sophistication of nucleic acid extraction, structural and functional analysis and synthesis techniques, development of cell operation techniques, development of individual production

Figure 3. **FORMULATION OF GOVERNMENT POLICY**

---

1. Administrative structure

a) *Council for Science and Technology* — the supreme deliberative body to the Prime Minister on science and technology policy. Panel on Life Sciences.

b) *Science and Technology Agency* — responsible for formulation of basic science and technology policy, general co-ordination of works by ministries and agencies concerned, and implementation of comprehensive R&D (except those for research at universities).

c) *Environment Agency* — responsible for environmental protection and related R&D.

d) *Ministry of Education* — responsible for university research.

e) *Ministry of Health and Welfare* — responsible for public health, medical treatment, medical products and related R&D.

f) *Ministry of Agriculture, Forestry and Fisheries* — responsible for agriculture, forestry and fisheries and related R&D.

g) *Ministry of International Trade and Industry* — responsible for mining, industry, energy and related R&D.

h) *Ministry of Construction* — responsible for sewerage and related R&D.

---

2. Publicized plans

a) General Guidelines for Science and Technology Policy (Decided by Cabinet) — life sciences is listed as one of the substantial R&D fields.

b) Comprehensive Fundamental Policy for Promotion of Science and Technology to Focus Current Changing Situations form the Long Term View (Recommendation 11 by the Council for Science and Technology) — life sciences is recognized as a leading and fundamental science and technology, and is discussed in biological and human aspects.

c) Basic Plan for Research and Development on Leading and Fundamental Technology in Life Sciences (Recommendation 10 by the Council for Science and Technology) — direction and enhancing measures for developing manipulation of genetic information systems are pointed.

d) 10-Year Strategic Program on Cancer Research (Decided by Ministerial Meeting for Cancer Control) — research themes for elucidating cancer and related measures.

---

techniques, and so forth, as well as a human series of scientific investigations to elucidate high-order functions peculiar to man, centring about the brain and nerve system and immuno-system. The Panel on Life Sciences is studying the last two subjects in details.

As regards gene and cell technology, the recommendation entitled "A Basic Plan for R&D of Pioneering and Basic Technology in the Life Sciences" was published in 1984. It pointed the direction and enhancing measures for developing manipulation of genetic information, emphasis being placed on research and development on chromosome level, new host-vector systems and effectively producing gene products.

Another aspect of the national plan arose from a ministerial meeting on a matter of specific significance. "The 10-year Strategic Program on Cancer Research" was decided by the Ministerial Meeting for Cancer Control, which is to foster research on cancer detection and on common basic technology, to strengthen exchange of researchers, and to provide research opportunities for young researchers.

Two further ministerial meetings, one for a better aging society and the other for AIDS control, have already announced their respective guidelines referring to the necessity of research and development.

IV. MAJOR POLICY MEASURES

Regarding the execution of the before mentioned basic policies, ministries and agencies concerned carry out research and development within their respective jurisdiction. If duplication of research and development arises among them, the Science and Technology Agency co-ordinates them during the course of the budget process. With this system, each ministry and agency is able to exhibit its policy planning ability and they compete with each other, especially in the field of basic research.

The Science and Technology Agency, on the other hand, promotes inter-ministerial R&D in line with the direction of the Council for Science and Technology. There are seven projects on life sciences: bio-membrane research on cellar level; structural and functional analysis of functional proteins, and their designing, production, and mimetic technology; supporting technology development for cancer research; measuring technology development for brain function analysis; comprehensive research on chromosomes; R&D on energy conversion systems of biological organs; and R&D on the aging process.

The agency also has two research systems: one is ERATO and the other is the International Frontier Research Programs.

ERATO is a unique system for the Japanese, where talented leaders recruit young researchers, and organize and conduct research teams for five years. The leader has an eye for appraising promising researchers and deep insight into his research topic. The research topic is provided only as a starting point and the leader has overall responsibility for the execution of his project. "Bioholonics" is to explore co-operative interaction between the individual elements (holons) making up a system, and the system itself. "Bio-information transfer" aims at elucidating the mechanisms of intercellar and intracellar information transfer; "Super Bugs" will search for and utilize micro-organisms under extreme environments such as strong pH, high temperature, salinity and pressure; and "Bio-photon" is to study the relationship between photons emitted or absorbed by biological organisms and cells are ongoing projects in the field of life sciences.

The International Frontier Research Programs, inaugurated last autumn, aim at actively discovering new knowledge that will serve as the basis of technological innovations in the 21st century, through gathering competent scientists under an internationally open system. The "Bio-homeostasis Research Program" is to elucidate the mechanism of regulating physiological function and maintaining its balance in animals and plants at four laboratories of Molecular Regulation of Aging, Aging Process, Intestinal Flora and Plant Biological Regulation.

Figure 4. **MAJOR POLICY MEASURES**

---

1. **Fostering Research and Development**

a) *Science and Technology Agency*

R&D with Special Coordination Funds for Promoting Science and Technology — inter-ministerial R&D under CST's supervision.

Exploratory Research for Advanced Technology (ERATO) — research system where talented leaders conduct teams of researchers from industrial, academic and governmental sectors for 5 years.

International Frontier Research Programs — research system where researchers from various circles gather for a specific subject such as biological background of homeostasis.

b) *Ministry of Education*

Grant for Scientific Research — comprehensive research projects are carried out.

c) *Ministry of Health and Welfare*

Basic Research on Health Sciences for Better Aging — joint research projects by national institutes, private laboratories and universities, and research supporting activities to assist these research projects.

d) *Ministry of Agriculture, Forestry and Fisheries*

Promotion of Breeding through the Application of Biotechnology for the Year 2000.

Projects for the Development of New Technology Related to the Food Industry, etc.

e) *Ministry of International Trade and Industry*

Research and Development on Basic Technologies for Future Industries — R&D on basic technology of the coming generation, such as bioreactors, large-scale cell cultivation, recombinant DNA and bio-electronic devices.

National Research and Development Program — R&D with particular importance and urgency from the national point of view, such as a water recycling system with a bioreactor.

---

2. **Fostering venture business and technology transfer**

a) *Science and Technology Agency*

Development of new technology and technology transfer — contract development with risk money and patent licensing.

b) *Ministry of Education*

Technology transfer — licensing of university-owned patents.

c) *Ministry of Health and Welfare*

Fostering venture business on medical products for basic health sciences.

d) *Ministry of Agriculture, Forestry and Fisheries*

Investment on Venture Business — fostering venture business on specified bio-industries.

e) *Ministry of International Trade and Industry*

Investment on venture business — fostering venture business on R&D of mining, industry and energy.

Technology transfer — licensing patents owned by AIST to industry.

The Ministry of Education is carrying out comprehensive basic research projects on life sciences based upon proposals from university scientists. The projects include cancer research, cell functions and calcium ions, plant breeding on cellar and molecular levels, and many others.

The Ministry of Health and Welfare promotes basic research on health sciences for better aging jointly among national institutes, private laboratories and universities on a project basis, with supporting activities.

The Ministry of Agriculture, Forestry and Fisheries is applying biotechnology to breeding, and accelerating technological innovation in the food, agricultural chemicals and fertilizer industries.

Bio-reactor, large scale cell cultivation, recombinant DNA technique and bio-electronic device are carried out as the projects of the Research and Development on Basic Technologies for Future Industries by the Ministry of International Trade and Industry. In its National Research and Development Program, the Ministry is also advancing a water recycling system with a bio-reactor.

The focus of governmental R&D in Japan is shifting to basic research.

Concerning the commercialization of research results from the public sector, the Research Development Corporation of Japan, under the Science and Technology Agency, the Japan Society for the Promotion of Science under the Ministry of Education, and the Japan Industrial Technology Association under the Ministry of International Trade and Industry have been taking the major role for years. Recently the need for set-up investment for venture business in basic research has been growing. Two organisations are in operation under the Ministry of International Trade and Industry and the Ministry of Agriculture, Forestry and Fisheries, and a new one under the Ministry of Health and Welfare will start this fall.

V. FUTURE ASPECTS

At this moment, the major direct policy objectives are strengthening international activities and enhancing basic research.

a) Strengthening international activities

Several measures have been taken on international exchange. As regards the national research institutes, the Science and Technology Agency has a scheme for sending researchers to overseas institutes, and inviting overseas researchers to Japanese national institutes. The STA plans to allocate additional budget for international exchange themes under governmental frameworks of agreement and arrangement for scientific and technological co-operation. In the academic fields, the Japan Society for the Promotion of Science has been taking the major role. These activities are general and are not focused on the field of life sciences.

The Council for Science and Technology and ministries and agencies concerned are now designing a plan called "Human Frontier Science Program". This is a large scale program aiming at a new realm of science through international basic research on the elucidation on the mechanisms of living organs, and in the course of such research a host of "seeds" and a variety of ripple effects will be obtained. Last week a feasibility study on research subjects by

Figure 5. **FUTURE ASPECTS**

---

1. Strengthening International Activities

— Human Frontier Science Program

2. Enhancing Basic Research

---

eminent scholars was finished, and the following seven research fields of importance and five leading key technologies were selected:

i) Perception and cognition: to elucidate the brain's skill process and the integration of information gathered through sense organs which are highly sensitive and densely packed.

ii) Motor and behavior control: to elucidate the neutral mechanism for free and complex voluntary movement which is based on instinct and emotion, and intelligently controlled to match the complex environmental and social conditions.

iii) Memory and learning: to elucidate the memory system that can store, systematize, retrieve and recall enormous amount of information with certain contexts, and the learning function through which one can acquire new behavioral patterns and functions, particularly in the development period.

iv) Expression of genetic information: to determine the full content of the genetic information expression mechanism, which is common to all organisms, from micro-organisms to man.

v) Morphogenesis: to elucidate the mechanism and adjustability through which complex forms and structures of living bodies are created.

vi Molecular recognition and responses: to explain the facts which are phenomenologically understood, on a molecular level.

vii) Energy conversion (including movement function): to elucidate the connection between molecular organization and the energy conversion mechanisms of various systems.

The five key technologies are: sequence analysis of DNA, analysis of the steric structure of proteins, non-invasive determination of biological functions, determination of the dynamic structure of biological systems, and ultramicro-manipulation.

Based on these promising themes, the administrative side is devoting its energies to devising a "best-fit" international scheme, recognizing the importance of providing well-established research environment for hopeful young researchers, forming international research teams, and establishing an infrastructure of scientific equipment such as DNA sequencers and data-bases on DNA, proteins, etc.

b) Enhancing basic research

In order to enhance basic research at national institutes, some Y 700 million was allocated from the Special Co-ordination Funds for Promoting Science and Technology in the 1986 fiscal year. Similar amount may be appropriated hereafter, and new initiatives by the respective ministries and agencies are anticipated.

This is an outline of the systems and aspects of Japanese R&D on life sciences.

The commercialisation of government and university research, and public acceptance of biotechnology

Dr. K. Heusler, Director of Research,
Ciba-Geigy AG, Basle, Switzerland

Biotechnology, defined as the use of biological systems to generate products and services, is well established in one or the other form, both in industrialised and non-industrialised countries.

The fermentation industry, whether it produces beverages such as beer or wine, or enzymes for laundry or other purposes, antibiotics, or glutamic acid as a food additive, or amino acids or artificial sweeteners, is an established industry. This industry relies mainly on two academic disciplines: microbiology and fermentation technology. Collaboration between universities and industry has been frequent: brewers had their own training centers where research was done. Most of the original antibiotics, the streptomycins, macrolides, tetracyclins, penicillins and cephalosporins were discovered in university or government laboratories.

Why are we now talking about potential or real problems in the commercial exploitation of university and government research? I think the reason for this is twofold:

1. The scientific development in molecular biology which culminated in the new technique of *genetic engineering* opened up a new dimension for biotechnology. By these methods not only the natural varieties of micro-organisms and cells were available for exploitation, but these organisms could be modified at will to perform new tasks which they would never have performed without this new technique. Opportunities seemed suddenly unlimited.
2. Since the method was novel, it was relatively easy for a skilled scientist to transgress the limits of classical biotechnology and express a gene product in an organism which could never have done this before. If the gene product was one relating to a system essential to a life process, the achievement was rightly called a break-through.

The new opportunities attracted much public excitement (and money!), but the risks and difficulties associated with the commercialization were much less obvious.

I. POLICIES TO FACILITATE UNIVERSITY/INDUSTRY CO-OPERATION

The factors critical to successful collaboration between two partners are a number of truisms derived from experience. However, these tend to become forgotten when the opportunities seem so marvellous that risks seem to be irrelevant:

1. Co-operation between two partners can only work and work successfully if there is a *common interest* in the subject which forms the basis of the common endeavour. To earn as much money as possible is not a sufficient common interest.
2. Co-operation must be *mutually beneficial*, both partners must gain something, e.g. access to knowledge or technology; here again money is not enough.
3. Co-operation also must be based on mutual *trust* and *recognition* of the strength and *capabilities* of each partner. This is especially true in scientific/technical co-operations because many details of the common work cannot be defined *a priori* and the size and importance of the contribution of each partner is unknown at the beginning.
4. Co-operation also crucially depends on communication. It is not surprising therefore that in all co-operation agreements, articles appear which define the channels and frequencies of communication.
5. Co-operation also requires a commitment of both partners.

At least a firm and irrevocable initial commitment of effort is essential. Agreements in which any of the partners can withdraw at any time on short notice are doomed to fail.

The different assignments and goals of universities and industry are shown in Figure 1.

Figure 1. **ASSIGNMENTS AND GOALS**

	Assignments	Goals
Universities	Education and Training Basic Research	Professional Experts New Knowledge Public Recognition
Industry	Applied Research Development Production Marketing	Commercial Processes Commercial Products Commercial Services

What significance do these prerequisites have for the topic of our discussion University-Industry Co-operation in biotechnology? Basically the assignments and goals of the two partners are very different. The classical assignment of universities is twofold: Education and Research. The public (including industry) expects that the university teachers and graduates are experts in their fields and that new knowledge is generated. For the university professor, recognition by his peers is an important goal.

For industry the tasks are applied research, development, production and marketing and its achievements are measures by the commercial success.

In view of these differences the question must be put forward: what would induce the two different organizations (university and industry) to co-operate? I have said already that the

Figure 2. **PROJECT COSTS (INVESTMENTS)**

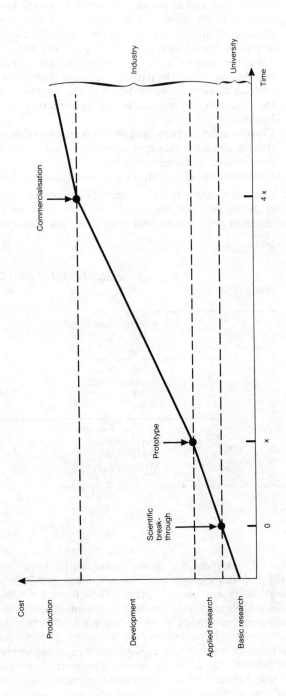

flow of money from industry to academia (or to an individual in the academic community) would not be a sufficient basis.

As shown in Figure 2 the investment risk in any area, but particularly in biotechnology is too much on the side of industry, both in terms of time and expenses.

What is needed are the common interests to achieve something that is socially relevant, scientifically rewarding and commercially viable. Then the project is mutually beneficial. The two partners will find each other if the academic partner brings unique knowledge and a very high level of expertise and if the industrial partner has the strength and experience to carry the project through to the end, if both are committed to the project and willing and able to communicate efficiently.

What can governments and public servants do in order to facilitate such co-operations? I have identified four points which seem to me particularly relevant. I am sure that there are more, but let us consider these four first.

In order to bring competitive advantages to an industrial partner the academic partner must be first rate, i.e. a department or institute of international reputation. Both multinational corporations and new biotechnology firms choose their academic partners irrespective of national borders. The most striking, and for many the most shocking example was the Hoechst-Massachusetts General Hospital deal some years ago. Governments must therefore strive at bringing at least *one* of their schools into the top league and maintain it as a real *center of excellence*. This argues strongly *against* spreading support to all schools who apply for biotechnology funds, but strongly *for* a careful, selected but continuous support of all the disciplines important for biotechnology at one or (in large countries) at several universities.

Although centers of excellence can in the great majority of cases be identified from the scientific and technical literature, small and medium size enterprises do not regularly have access to these sources. Government agencies must therefore provide *information*, but this information should not be exclusively national.

It is usually difficult enough to formulate the conditions in a co-operation agreement to serve the interests of both partners, for example to satisfy the legitimate desire of the academic partner for publication and free communication and the industry's wish for secrecy. Usually a compromise can be found, but the solutions can be facilitated by special legal provisions:

— Patentability of biotechnology inventions does not prevent communication, because the correlate of the privilege granted is public disclosure;
— A worldwide introduction of a "grace period" for patent applications by the original inventor facilitating early publication; and
— The removal of conditions which forbid exclusive licensing of inventions which involved support by public money increases the attractiveness of co-operation.

It must be realized, first, that the costs involved in developing an invention into a commercial product or process are such that they can only be recovered if exclusive and preferably worldwide licenses can be obtained for a considerable period of time. Secondly, the efforts which are necessarily involved in co-ordination among several licencees are considerable and, in addition, the risks of violation of antitrust laws in cases of multiple licenses are not to be neglected. To my knowledge the US Department of Justice has therefore loosened the restrictions for licensing of inventions financed or co-financed with federal agencies' support.

Finally governments can create incentives which favour R&D investments in general, such as tax incentives, a stable economic climate and clear and straight-forward rules and regulations which guarantee the partners that they are not faced with new and unforeseen regulatory difficulties down the road in their collaboration. Last of all I mention financial support of collaborative efforts only because in most cases viable projects take off without state support and if the availability of this support is the only real reason for the initiation of a co-operative project its chances of success are very often low.

Some concern has been expressed that the involvement of academic personnel in intimate collaboration with profit oriented organizations, sometimes even in management and board positions, had a detrimental effect on free scientific communication. While in some cases this has undeniably been true, it is in my opinion no longer a serious concern, because for those who want to stay in academia such behaviour would damage their scientific reputation, and the period when the short term financial rewards for an individual were very substantial, is gone.

II. PUBLIC ACCEPTANCE OF BIOTECHNOLOGY

If the opportunities of a new technique are so enormous, why should the public be afraid of them? There are again two reasons:

— Anything which is unknown is, *a priori*, frightening; and
— Modern biotechnology is in its concept and its executions extremely complex and, therefore, difficult to grasp and understand by lay-men.

Background

A few years ago I was convinced that the fears and anxieties which were associated with biotechnology in public audiences were due to a lack of information; that the scientists had been unable to explain the background, the present state and the limitations of their new discoveries to the ordinary citizen in clear and understandable terms. Therefore it seemed reasonable to make every effort to correct this situation by an intensive information campaign which would explain in simple and attractive terms to the "man in the street" the principles, as well as the achievements of biotechnology. It seemed entirely possible that, on the basis of such information, he would clearly see the unique opportunities which this new technology offers and that the potential dangers are relatively minor compared to the potential benefits.

In the last few years a number of first class information campaigns, television programs, tapes and books on biotechnology have been produced. The television series in Holland, Great Britain and Germany (ZDF) are well done and accurate. The German programs are being shown in Switzerland and Austria. More and more people are informed, but the fears and anxieties have not disappeared. On the contrary, if anything, they have increased. Why?

Sociologists have long ago realized that over-information generates cognitive stress. The amount of information necessary to grasp the importance of biotechnology requires an intellectual effort which not many people are willing to invest. The German television series consists of 13 chapters of approximately 45 minutes' duration each. It is also my impression that at the outset the benefits of biotechnology were oversold. The relatively slow realization let the glamour somewhat fade away and allowed the fears about the risks to grow steadily. In order to avoid the stress of over-information it is much simpler to follow those who simply say that anything new which cannot be understood, and about which not even the experts can exactly and simply say where it will lead us, should be rejected: "The world is bad enough, let's not make it worse".

It is true that any new discoveries carry with them a certain amount of uncertainty. This was also true at the time of the discovery of new continents. The anxiety associated with uncertainty was clearly expressed by a man who drew a map of the world at that time and wrote in the areas of the world which had not yet been explored by man: "Here are the dragons". But although the explorers were frightened by them, nothing would prevent them from moving forward and fighting the dragons if they should show up. Today it is oftentimes assumed that the unknown is not only dangerous but overall it can only be bad and destructive. Therefore it is not even justified to explore the unknown.

In such circumstances factual information is no longer in demand. Any information which deviates from one's own preconceived opinion is classified as manipulated information or ignored. The frequency with which a particular position is reiterated determines its weight, the arguments on which it is based become unimportant.

There is a second development which is a reflexion of the same attitude: the discrepancy between *acceptability and acceptance*. Even if, on the basis of rational arguments and the careful evaluation of the benefits and risks associated with a particular project or action, social partners can agree that the project or action is perfectly acceptable, particularly in comparison with alternative solutions, this does not mean that both partners publicly accept the project or action. Although rationally the partners agreed, they did not trust each other because there might have been facts or issues which had not surfaced during the evaluation.

Acceptability and acceptance are not the same thing. Acceptability is based on the rational evaluation of benefits versus risks. The acceptability must be continuously reassessed on the basis of the totality of information available at any given time.

Acceptance, however, is an emotional judgement based on feelings or beliefs of social and ethical responsibility. Acceptance is mostly linked to trust.

Possible actions

In the documentation for this meeting includes a statement that "the acceptance of biotechnological products and processes by the public will have a significant impact on the pace of commercialization in these fields. Questions of safety and ethics are obscuring the benefits, both economic and social, that will be derived from the application of biotechnology (...). Public awareness programmes will be an important component of any strategy to encourage biotechnology development".

We can certainly not deny the importance of public acceptance because we all depend on it. Governments cannot dissociate themselves from public opinion: legislation (new rules and regulations) is influenced by both risk assessment and risk acceptance by the public. Academic research depends on the amount of funds made available by governments and their decisions are influenced by politicians and public opinion. And, last but not least, industry is only willing to develop biotechnological products and processes if the chances of getting approval and the requirements for approval are reasonable and economically acceptable. In a recent paper in "Risk Analysis" Chauncey Starr[1] concludes that "Public acceptance of any risk is more dependent on public confidence in risk management than on the quantitative estimates of risk consequences, probabilities and magnitudes".

What should therefore be the content of a public awareness program to increase public acceptance of biotechnology? From what we have said before it cannot be more information on all the wonderful things biotechnology will be able to do. One approach which would increase the confidence level would in fact be risk management. Even with this approach we are faced with formidable opposition. In the Founding Appeal of the international Gen-Ethic Network (a non-profit organization incorporated in West Berlin) we read:

"Today it is possible to set into motion global processes which can neither be reversed nor reliably restrained, and which may end in the annihilation of our planet. It is the ruling paradigm of scientific and technical "progress" which has brought us to this level of potential destruction — a paradigm which dictates that whatever can be done *must* be done, and that it is only in the application of new knowledge that the consequences become evident. This notion of unrestrained "progress" based on trial-and-error is clearly no longer adequate to the discussion of the problems we face today. We must develop a new ethic for dealing with our knowledge and cannot entrust this task only to scientitist, politicians and so-called experts, nor can we leave it to the mechanisms of the free market and international competition".

I think it is our task to demonstrate that by by proper measures the process *can* be reliably restrained and that "progress" towards new horizons is possible in a careful and controlled manner. This is what we mean by risk management. But we muste be able to put our own position forward in a credible way in order to achieve public acceptance. This is not going to be easy. Perhaps we can derive some guidance from an article published by H.C. Roeglin[2] develops four principles for public relations work in this context:

1. The credibility of the informant is a decisive component.
2. Overinformation leads to passive refusal of one-way information. People should be offered the opportunity to inform themselves, information should not be forced upon them. Do not answer questions which have never been asked.
3. There are always two sides of a coin: there are benefits as well as risks. Acceptance of benefits persupposes exposition of the associated risks. It is wrong to start from the assumption: let sleeping dogs sleep: the dogs wide awake.
4. Information should not preempt decisions. It should make facts transparent, encourage rational evaluation and facilitate decisions.

Ad 1.   In order to be credible, the informant should not profit in any way nor suffer unduly from a decision in the area in question; he should be informed but also be aware of the concerns of the public. He should be able to express himself clearly. Such persons are difficult to find.

Ad 2.   It is important to keep track of the questions which really are in the minds of people. It does not serve any purpose if the recipient of the information either already knows what we are telling him of if he does not care about the information. Let him ask the questions.

Ad 3.   Nothing is without risk: therefore expose the risks, then explain the potential benefits. Let the listener himsel decide whether the benefits outweigh the risks.

Ad 4.   It is also of little help to point out that other things or actions are much more dangerous than what you would like to propose. Each case must be judged on its own merits.

In order to increase public public acceptance at this time it might be more important to cite examples of projects which were abandoned because they did not fulfil expectations or where the risks were becoming more important than the visible benefits.

Optimistic success stories are no longer in demand, especially if they are projected into the future, even though this behaviour might have negative effects ont the stock value of some companies. In the long term such behaviour will be honoured by the public and without the acceptance by the majority of the citizens the uncertainties of any action are just too high.

These are some thoughts on topics which will be further discussed during our meeting. I hope that they have given some indications of the directions into which our recommendations should go.

REFERENCES

1. Starr, Ch. *Risk Analysis 5*, 1985, pp. 90-102.
2. Roeglin, H.C., *gdi impuls, publication of the "G. Duttweiler Institut für Entscheidungsträger in Wirtschaft und Gesellschaft"*, Rüschlikon, Switzerland, 1, 1985, pp. 61-68.

Training and mobilisation of human resources and the flow of scientific information

Prof. Daniel Thomas, University of Compiègne,
Director of the "Programme mobilisateur des biotechnologies"
Ministry of Research and Higher Education, France

This report will discuss training, including all the training and manpower resource problems encountered in biotechnology, and the free movement of scientific information and scientific data throughout the world. The discussion on training will proceed in three stages. First there will be a general account on the interaction between training and research. This will be followed by a description of certain scientific fields where training is evolving very rapidly. Finally the problem of human resources, in broad terms the interaction between training in biotechnology, production and industrial activity, will be addressed.

The first main consideration arises from the fact that some authors have tried to argue that there is only one kind of biotechnologist, trained according to one single pattern and able to work effectively in all situations. No biotechnologist is capable of meeting all situations. There must be different types of training. Suffice it to say that those trained in biotechnology do not all have the same profile but must have received sound training in a discipline such as microbiology, molecular biology, biochemical engineering and purification methods and must also have broader knowledge of other fields. The most useful type of training should therefore provide sound knowledge of one field combined with discerning knowledge of all the other fields. However in biology knowledge has evolved extremely rapidly. There has been a qualitative leap and the training of those involved in biotechnology must therefore be conducted in close relation with research.

The link with research covers both training for research and training through contact with research. In our respective countries and mine in particular, some types of training are too far removed from research and, especially, high standard research. Particularly in this field, no training can be provided in higher education if there are no contacts with research of extremely high standard.

The other fundamental feature shown by biotechnology is multidisciplinarity mainly bringing together two major sciences, biology and engineering. Biotechnology is of interest for engineering sciences not only because it is applied but also because through biotechnology, there has been a change in the scientific approach used in biology. Until recently, this approach chiefly consisted in discovering and understanding what was going on in nature.

Henceforth, it consists in creating, inventing and innovating and biology can therefore now be incorporated in the intellectual scientific approach used by engineers. Biotechnology training is very important and must be integrated in the scientific and technical culture of today's engineers. It is no longer possible for any engineer in any field not to have minimum knowledge of biology and biotechnology.

This general view of training and relations with research shows that the relationship is dynamic, that by evolving research modifies the training content and also that biology forms part of training in engineering sciences.

At this stage it is worth looking at a few examples where things have evolved. Six scientific fields where major training changes have occurred illustrate how difficult it is to select the right training. The target aimed at is mobile and it is very difficult to predict the requirements of research and industry in five years' time.

The first of these six fields is molecular biology and genetic engineering, which used to be regarded as a limiting factor. Even a few years ago, demand was high and considerable efforts had to be made in training. In France for instance, much work was done to introduce molecular biology and genetic engineering training. To some extent, too many efforts were deployed and, possibly more to the point, they were not put sufficiently into perspective. This is because molecular biology and genetic engineering are essential for the implementation of current programmes, but seldom act as the limiting factor. Training courses must be adjusted to foreseeable trends in limiting factors for industrial programmes. In our countries, it is now no longer sufficient to have received training in genetic engineering only. Genetic engineering must be seen in perspective with the other disciplines and trends in technology and industry.

In the second scientific field, microbiology, as the Swedish Delegation pointed out an effort must be made to give the associated disciplines their rightful place. They are becoming less attractive to scientists and in future, there might not be any Nobel prizes in microbiology. Yet microbiology forms part of biotechnology training and without high standards in microbiology, there cannot be any biotechnology. Tribute should be made to our Japanese friends, who have constantly succeeded in maintaining very high standards in microbiology training, which is not always the case in other countries. In mine for instance, I feel the situation is not entirely satisfactory.

The third field is that of proteins. Scientific interest is currently shifting away from DNA to proteins, partly owing to the spectacular results achieved in the DNA field. So much progress has been made in the genetic programming of functions that the limiting factor from the scientific, technological and industrial standpoint is now knowledge of the function, in other words the protein. This requires knowledge in physics for the three dimensional structure of proteins, in computer processing for molecular graphics, expert systems, artificial intelligence and of course, the use of the genetic engineering tool of controlled mutagenesis.

Proteins are now becoming an important aspect. Clearly we should have foreseen the introduction of many more training courses on proteins including all aspects of extraction and purification. One of the limitations in our countries at the moment is the absence of purification and extraction specialists. The success of competing industrial programmes often depends on protein and product purification using increasingly sophisticated methods.

The fourth scientific field concerns fermentation and biochemical engineering, which in the United States for instance, had been slightly neglected lately but has now come back to the fore. In this field, highly interdisciplinary research is conducted, covering microbial physiology, chemical engineering (decision tool), sensors and biosensors, an essential area in which training must be given, and finally the use of computer control in real time and the most modern methods of automation.

The fifth field is molecular biology and the plant world. Training must incorporate molecular biology in the plant world but without underestimating the huge scientific legacy of conventional approaches. Nowadays, effective training in this field must be both alert to developments in the DNA field and transfer methods and fully informed of conventional plant improvement methods and all selection methods. This cross-fertilization is a major requirement for effective training in this field.

Finally, immunology has remained too long the prerogative of medicine but it is now literally exploding since it covers the veterinary field, plants, agriculture, agro-foods, cosmetics, perfumes, the environment, etc., in other words a very wide variety of areas. Immunology taught from a strictly medical standpoint is now of much less interest than in the past and arrangements must be made for immunological concepts to be taught in almost all biotechnology areas. The possibility of using this tool in fields even far removed from medicine and, of course, at industrial level must also be taken into account in training.

So far, after a brief discussion, we have looked at six scientific fields where training problems are being experienced. I should now like to comment on human resources in the broad sense. Biotechnology undoubtedly covers a number of advanced scientific and technological disciplines. But these do not only apply to new advanced industrial activities but also to highly traditional industrial ones. In this context, not only must a very high standard of research-linked training be maintained but the initial and continuous training of those involved in biotechnology including engineers, technicians and perhaps even manual workers employed in the different industrial activities concerned by biotechnology must also be taken into account. This is an important point because simply bringing in highly qualified scientists from outside will not be sufficient to solve all training problems. The basic biotechnological knowledge of almost all agents in the agricultural and industrial use of biotechnology must be increased. This will be difficult and may take a long time to achieve but it is essential for all actors.

These aspects of training have different components, the first one being fundamental changes in the nature of biology in relation to the development of production processes. Biology is becoming a process of creating, innovating and inventing new functions and not just merely discovery. This evolution is very important. It must be taken into account in the training of engineers and not just in biotechnology as such.

When giving these six examples of scientific disciplines, I have attempted to show that it is very important to be alert to all developments in science and industry in order to predict future changes. At the moment, anyone strictly specialised in DNA chemistry is experiencing great difficulties in adjusting to these changes. We must therefore try to predict where the target lies and we must ensure flexible and evolving training. This can only be achieved through contact with research of the highest standard. Nowadays research must be of use to biotechnology and of high standard particularly if it is required for training purposes.

After dealing with training issues we shall now discuss the free movement of scientific training, ideas and biological material.

Firstly, as reaffirmed by several delegations including the United States, university and academic centres must continue to conduct basic research defined in relation to their own strategy and reaching the highest standards. The contacts between university and industry concerning basic academic research must be extensive while remaining in line with the advancement of knowledge in basic university research. It would be extremely dangerous in national programmes to divert part of the basic research from these objectives towards very short-term technological ones. The technological aspects must be developed in contact with academic research of extremely high standard.

The problems of the movement of scientific information will be mentioned in connection with six areas. The first concerns conferences, participation in conferences and the attendance of highly qualified scientists. Recently we have seen conferences actually concerned with exchanges of scientific information and activities relating to academic research and the transfer of information and knowledge, but there have also been too many symposia, frequently bearing the title "bio something", which were more akin to commercial events than venues where scientific information could actually be transferred. We must watch events closely because the impression might be gained that there are many opportunities for comparing notes. In fact, there are relatively few venues where information is actually exchanged.

The second important area is that of the scientific press, journals and the dissemination of findings. Here also we must watch events closely, because it is important that basic knowledge should not be privatised too early and should be disseminated broadly through the scientific press. In some cases, special links have been created between major university laboratories and industrial firms, sometimes leading to a kind of privatisation of fundamental knowledge. Open competition must occur on the market in biotechnology and it must be free competition. However, fundamental knowledge must continue to develop at its own pace and it might be extremely dangerous to limit it through rampant privatisation. In the 1982 report, we already emphasized the fact that contacts between university and industry should be extended while preventing any privatisation of fundamental knowledge which would result in limiting the movement of information and, in the final instance, the dissemination of information in all journals and books and all the conventional supports for biology and biotechnology.

The third problem is that of the exchange of persons. It is abundantly clear that compared with reading the literature and attending meetings, a far more effective method for inducing the transfer of information is the movement of researchers and scientists.

The best possible vector for the transfer of scientific information is the training of individuals and their transfer from one organisation to another. One of the difficulties arises from the competition not just between firms but also between countries, leading to restrictions affecting the movement of scientists, and care must be taken to ensure that not just ideas but also persons can move between the different structures. The departure of a well-trained scientist seeking different experience in another country can be construed as a loss for his country of origin. In their meetings, international organisations and the different countries should be aware of such problems and reflect on the perfectly legitimate possibility of safeguarding their interests while allowing the movement of persons between countries so as to ensure that evolving science, the science of the future and not just dead science, can circulate among our countries.

The fourth area concerns databanks, which are of great importance. It is no longer possible to work in biotechnology without databanks and without adequate access to databanks. Clearly a problem arises with the location of databanks and access to them. Countries, international organisations and groups of countries such as the European Community must obviously study this type of problem because access to such databanks must be as broad as possible, thereby creating a situation where competition between enterprises is completely loyal since scientific information can be obtained through the databanks from anywhere in the world. The problem of databanks is already largely settled but much effort is still required concerning standardization, co-operation and synergy between the different bodies worldwide.

The next problem concerns the movement of living material and hence collections of micro-organism and plant strains. Some collections are widely accessible but co-ordination is required, given that in biology free movement of information entails free movement of biological material since the latter is involved in the basic scientific approach. The control of industrial property is a separate issue which is currently being discussed.

The final problem concerns industrial property. Free movement of information is closely governed by the nature of industrial property. The requirements of industry may in certain cases complement those of the academic field. Provided industrial protection is effected under satisfactory conditions of transparency, better dissemination of information will ensue since there will be fewer constraints.

In the case of patents, the grace period must be taken into account as it is a means of facilitating the dissemination and transfer of information. On the whole the scientific community views the grace period very favourably. Some countries are generally in favour but others, such as mine, have misgivings. We are facing a dilemma: although the grace period raises a few difficulties relating to conventional industrial property it opens extremely important and positive horizons for information transfer. The entire scientific community is very aware of this. The difficulties which may arise in some countries can be overcome only by taking into account the overall situation concerning all international problems, looking at all international trends at a given moment in time and not by confining it to minor local aspects possibly having an impact on a few specific interests.

In dealing with the movement of scientific knowledge and ideas, I attempted to discuss six problems which seemed important.

In conclusion, on the whole the two aspects I have described are not at all separate. Clearly the movement of scientific information is an extremely important requirement in training. Furthermore, training as many scientists as possible is one of the means of ensuring the movement of scientific information.

Some of the points requiring further investigation undoubtedly include improved harmonization of advances both in research and technology and in training. There must be contacts between training and research but it is not always easy to find practical and effective ways of doing this.

The other aspect concerning training is the successful implantation of technology throughout industry, which depends on improving the biotechnological knowledge of almost all actors in industry, be they manual workers, technicians or engineers.

As far as the free movement of information is concerned, if one aspect is to be emphasized it could be the free circulation of persons between countries and between academic and industrial bodies as we have undoubtedly reached such a high level of technical complexity that it is not always enough merely to read the literature. To ensure the flow of scientific knowledge from one structure to another the persons who are engaged in scientific activities must be able to move under optimum conditions.

The problem of training and information in biotechnology is extremely difficult but if it is not solved, the future of biotechnology may well be in danger.

Safety and regulations, R&D and international co-operation in biotechnology

Dr. David T. Kingsbury, Assistant Director for Biological, Behavioral and Social Sciences, National Sciences Foundation, United States

On June 26, 1986 the United States Government published the final part of the "Co-ordinated Framework for the Regulation of Biotechnology" in the *Federal Register* (FR 51, N° 123, pp. 23302-23393). Our goal in developing the "Co-ordinated Framework" was to explain to the American Public that, for questions involving the products of "biotechnology", human health and the health of the environment were adequately protected. The policy guidelines are based on generally accepted scientific principles and, therefore, provide a rational, yet stringent, basis for regulation. We also sought to explain to the rest of the world our desire to "promote international scientific co-operation and understanding of scientific considerations in biotechnology ...". Likewise, we stated that "The United States also seeks to reduce barriers to international trade. US agencies apply the same regulation and approval procedures on domestic and foreign biotechnological products. We are seeking recognition among nations of the need to harmonize, to the maximum extent possible, national regulatory oversight activities concerning biotechnology. Barriers to trade on biotechnological products should be avoided as nations join together in working toward this mutual goal".

The commercialization of modern biotechnologies has been in progress for slightly over a decade. The initial products were monoclonal antibodies and new drugs that resulted from the large scale fermentation of micro-organisms which had been genetically modified. In none of these instances was the product the organisms themselves, but they were simply the means of production of the end product. Production in these cases was done in carefully controlled and contained facilities. As the industry continued to grow and expand into non-medical applications, it became clear that many of the products of the new industry were going to be the *living organisms themselves* and that most would be applied in the environment. This recognition led to a wave of concern regarding the environmental consequences of the application of these products. Those who have expressed the most concern have focused on the possible negative environmental impact of the new products and have generally disregarded the potential positive environmental impact from, for example, the replacement of toxic chemicals and pesticides with new microbial products. Our governments have a heavy stake in this problem. Modern biotechnology is a direct product of many years of government investment in biological and biomedical research. We are now beginning to reap the benefits of that research. The promise of biotechnology is starting to be realized. However, we must be vigilant to ensure that this past investment continues to pay dividends. One factor, that will be necessary for many years to come, is a continued commitment to research, both basic research in biological processes such as gene regulation, genetics, cell biology and plant sciences, and in the environmental sciences such as ecology, population biology and systematics. We are entering an era when the pharmaceutical and chemical industries are calling for a broader research agenda in generic scale-up production technology which will enable them to more quickly and economically produce their products. I would argue that it is equally important that we commit an adequate level of research funding to the environmental sciences in order to continue to build the broad data base necessary to ensure the intelligent and *scientifically based* regulatory decisions needed to enhance progress in the future.

We have begun the planning for one small part of this effort. We have proposed the establishment of an international data base related to the introduction of new organisms into the environment. The initial planning has been a joint activity of the US and the Commission of the European Communities. It is our intention to extend the activity outside of this group to include at least all of the Member countries of the OECD. During our initial planning workshop we developed a working model which consisted of a data base of seven interlinked data files. The seven files included: taxonomy; literature; organisms; release events; guidelines; a directory of related information sources; and, finally, an "electronic bulletin board" through which investigators and regulators could share ideas and data.

Within the taxonomy file we propose to provide the latest in taxonomic data for all species that are included in some form in all of the other data base files. In the literature file we propose a bibliographic file containing all of the references cited in any part of the other files. This file will be searchable by conventional means such as author, title, journal and any textword present in the abstract. In the organism file we propose an extensive synopses of known data, extracted from the literature by a technical staff and reviewed and scored for quality by an expert panel. Where appropriate, synopses are to be referenced with pointers to the literature file. A wide array of data about the biology, genetic description, uses and known genetic modifications will be included as will comments about the availability and source of the organism.

The release event file will consist of synopses of results of previously conducted pre-release and release experiments, extracted from the literature and from non-published technical reports. This file will include all of the known reports of the introduction of new organisms that have occurred over the past one hundred years.

The guideline file is intended to assist in the design of future experiments, providing information on relevant procedures, technical considerations and regulatory guidelines in effect in various countries or locales. The directory of related sources will reference such sources as GENBANK and the Protein Sequence Data Bank.

In discussing this concept we agreed that the amount of available data that have resulted from the past century of research on microbial ecology, soil ecology and chemistry and plant and animal breeding is enormous. The volume is so great that it has, in itself, been a barrier to obtaining comprehensive reviews, thereby, giving the impression that data do not exist. We hope that this suggested solution will address that problem and simultaneously assist those responsible for funding basic research and those responsible for research in support of regulation, to focus on the areas of greatest research need.

A survey of the world's literature on the introduction of new plant varieties and the use of micro-organisms in the environment immediately identifies the need for broad international participation in this data resource. Many countries of the world have had a much broader experience with the agricultural applications of micro-organisms than we have had in the United States and those experiences will aid us greatly as we try to examine future research directions. Likewise, such long term ecological studies as the Long Term Ecological Research (LTER) sites supported by the US National Science Foundation will be critical to the provision of baseline information upon which we can build interpretations of the possible immediate environmental perturbations associated with a large scale release of a new organism into the environment. We must remember that understanding ecological balance is based on longitudinal observation and cannot be represented by cross sectional "snap-shot" data collection. We have high expectations for the positive effect of having a central large data resource and call on each of you to help us establish and operate this activity.

While I have been focusing on the information needs of the scientific and regulatory community, we must pay equal attention to the information requirements of the public we serve. Public apprehension about some of the products of biotechnology has surfaced as a problem in many countries of the world. This public apprehension has been responsible for considerable delays in the field testing of new products, despite the fact that these small scale tests have been approved by the regulatory agencies. What can we do to assuage the fears of that small number of people who feel that this technology is a threat to the environment?

I believe that there are several approaches. Let me suggest two than can be realistically implemented. First, we must clarify the use of the term "biotechnology". I have spoken out in the past as I do here suggesting that this word has now become a significant millstone around the neck of both the industry and the government. Biotechnology is not a unitary entity, but instead is an enabling technology; it has broad applications in many diverse aspects of industry and commerce. As we use the term today biotechnology is the establishment of hybridomas for the production of monoclonal antibodies to be used in diagnostic kits or therapy. Likewise, it is the use of recombinant DNA technology for the production of a hepatitis B vaccine in yeast, the production of Interleukin II in *Escherichia coli* or the introduction of increased levels or higher nutritional value of storage protein in soybeans. Moreover, the use of recombinant DNA technology for the engineering of new microbial pesticides or microbes for ore recovery will be important future products. However, as the success of this technology has grown so has the interest of government leaders and the financial community. The result of this special focus has been the gradual broadening of the definition of "biotechnology" to include a number of techniques which have been used for decades, without the specific attention they now receive. Products derived from chemical or ultraviolet light mutagenesis, hybrid plants, and micro-organisms produced through traditional genetic exchange are often considered objects which should be subject to new levels of "biotechnology" related regulation. We have experienced this change as a result of several factors, but I believe the principal factor is our use of a single, imprecise term to describe those activities and that has raised public concern. We must find a means whereby we can describe the products to be regulated in light of the specific properties which led to their examination and not simply apply the term biotechnology. Past practice has led us to approach a process based regulatory matrix and to mislead the public about the safety of the products being produced. In the United States we have tried to avoid the process bases approach and our regulators are ever vigilant, however, some of our government leaders do not understand the technology of the new biology and have vague concerns based on poor understanding.

The second area of public information that is necessary is to find a means of telling the public about the accomplishments of the new biology and the already impressive list of products and reagents. Most of these are related to health and many more are soon to appear. Commissioner Young of the US Food and Drug Administration has taken an active role in attempting to make wide public information available when exciting new products are approved for sale. This is a good beginning, however, it is not enough by itself. The public must understand that the potential applications of "biotechnology" affect many parts of their lives and are not limited to the area of pesticides and protection from ice crystal formation. In addition to building this basis of confidence on good information about past accomplishments, *realistic assessments* about the future applications are critical. We must avoid creating excessive expectations, but at the same time the public should be aware that new microbial products to replace toxic chemical pesticides are possible, and imminent, if we keep the climate open for innovation. The application of these new technologies to other environmental problems must be pursued as a means to solve some of the industrialized world's most pressing problems. In the lesser developed countries new approaches to plant and animal agriculture will help overcome the pressing problems associated with poor distribution systems. We must somehow find ways to initiate information dissemination so that these technologies are appropriately exploited. One critical element of this is public confidence that we are not trading one environmental problem for another. This is why the US has worked diligently to put in place a sound regulatory policy and is constantly reviewing that policy in the context of new scientific findings. The challenge is to always be on the boundary between over regulation of new products with the accompanying repression of innovation, and under regulation with its potential for environmental harm, without crossing significantly into either side of that boundary.

A broad range of research activity related to assessing the environmental impacts of new products is essential. In the US we attempt to achieve that broad range of activity principally through the research programs of two government agencies. The combined programs of the Environmental Protection Agency and the National Science Foundation cover the full spectrum from theoretical ecology through specific ecosystem models for laboratory assessment of environmental effects. Broadly speaking, the National Science Foundation programs focus on the very basic side of this research spectrum with little direct attention paid to questions related to regulation, while the Environmental Protection Agency programs focus on many aspects of the regulatory needs, in addition to some fundamental work. In the middle area there are NSF grantees working together with EPA scientists on common approaches.

One example of this type of activity is represented by a workshop currently in the planning stages led by the Environmental Protection Agency with the help of the Biotechnology Science Co-ordinating Committee. The focus of this workshop is methodology for the cataloging, identification and enumeration of micro-organisms in the environment. Additional workshops are planned for the future which will cover a broad range of topics related to environmental microbiology.

We believe that this integrated activity will generate new knowledge and serve both the producers of new products for environmental application and the regulators who will assess the appropriateness of those products. We hope to reach a time when many of the basic organisms and genetic manipulations are characterised such that the strict case-by-case assessment currently in place can be transformed into categories which, as called for in the 1986 OECD document "Recombinant DNA Safety Considerations", will allow various classes of proposals to be excluded.

In summary I would like to emphasize the continued need for a broad and vigorous base of basic research with a balance between the fundamental work which may eventually lead to commercial products and the fundamental work necessary for us to understand the interaction of newly introduced organisms with the environment. I would like to reiterate the need for balance in the regulatory approach so that we do not repress innovation in research and development. Let us not forget that over regulation has many side effects. In addition to repressing innovation and not taking advantage of our research base, over regulation leads to reluctance by the capital markets to invest in the future of our new industries thereby halting their development at an early stage. Under regulation at the same time leads to lack of confidence by the public and paralysis of the industry based on public outcry and legal proceedings.

It is my personal belief that the combination of a sound approach to regulatory practice, based on current scientific knowledge, combined with the appropriate communication with the public regarding the new products will lead to an exciting future for all sectors of industry which use the new biology.

Annex

LIST OF PARTICIPANTS

Workshop Chairman

John R. Evans
Chairman and Chief Executive Officer
Allelix, Inc.
Mississauga, Ontario
Canada

Host Country—Canada

The Honourable Frank Oberle
Minister of State for Science
and Technology
Ottawa, Ontario

The Honourable Elmer MacKay
Minister of National Revenue
Ottawa, Ontario

Reiner Hollbach
Deputy Secretary
Industry, Trade and Technology Sector
Ministry of State for Science and
Technology
Ottawa, Ontario

Henri Rothschild
Director General
Ministry of State for Science and
Technology
Ottawa, Ontario

Laird Roe
Analyst, Biotechnology
Strategic Technologies Branch
Ministry of State for Science and
Technology
Ottawa, Ontario

Bruce Stuart
Communications Branch
Ministry of State for Science and
Technology
Ottawa, Ontario

Bruce Howe
Secretary and Chief Science Advisor
Ministry of State for Science and
Technology
Ottawa, Ontario

David Shindler
Manager
Biotechnology Unit
Ministry of State for Science and
Technology
Ottawa, Ontario

Janet Ferguson
University Research and
Granting Councils Branch
Ministry of State for Science and
Technology
Ottawa, Ontario

Keith Bailey
Director
Bureau of Drug Research
Health and Welfare Canada
Ottawa, Ontario

Barbara Craig
Secretary
Biotechnology Unit
Ministry of State for Science and
Technology
Ottawa, Ontario

David Carlisle
Research Advisor and Chief Scientist
Water Quality Branch
Conservation and Protection Service
Environment Canada
Hull, Quebec

Gilles Julien
Executive Director
Natural Sciences and
Engineering Research Council
Ottawa, Ontario

Brigitte Leger
Conseiller (Sciences et Technologie)
Mission du Canada auprès des
Communautés Européennes
Brussels
Belgium

M. Louise McArthur
Strategic Grants Officer
Natural Sciences and
Engineering Research Council
Centre
Department of Chemical Engineering
Ottawa, Ontario

Doug Paterson
Science, Technology and Communications
Division, External Affairs Canada
Ottawa, Ontario

Martin Boddington
Chief
Biotechnology Centre
Commercial Chemicals Branch
Conservation and Protection Service
Environment Canada
Hull, Quebec

Maurice Brossard
Vice-President, Biotechnology
National Research Council
Ottawa, Ontario

Ian de la Roche
Director General
Priorities and Strategies
Research Branch, Agriculture
Agriculture
Ottawa, Ontario

John Langstaff
Director of Research
ABI Biotechnology Inc.
Winnipeg, Manitoba

Don Lush
Beak Consultants Ltd.
Mississauga, Ontario

J.J. McGonigal
Director General
Food and Consumer Industries Branch
Department of Regional Industrial
Expansion
Ottawa, Ontario

Murray Moo-Young
Director, Industrial Biotechnology
University of Waterloo
Waterloo, Ontario

Other Participants

Denis Dewez
Scientific Counsellor
Embassy of Belgium
Ottawa, Ontario
Canada

Mark Cantley
DG XII, Concertation Unit for
Biotechnology in Europe (CUBE)
Commission of the European Communities
Brussels, Belgium

Peder Olesen Larsen
Chairman
Danish Council for Research Policy
and Planning
c/o Danish Research Administration
Copenhagen, Denmark

Paula Nybergh
TEKES
Helsinki, Finland

Daniel Thomas
Laboratoire de Technologie Enzymatique
Université de Compiègne
Centre de Recherche
Compiègne, France

Gerhard Siewert
Schering AG
Berlin, Germany

Michele Lener
Ministero per la Ricera Scientifica
e Technologica
Lungotevere T. de Revel
Rome
Italy

Shin Aoyama
Life Science Division
Research and Development Bureau, STA
Chiyoda-Ku, Tokyo
Japan

Rintaro Ishiwata
Managing Director
Sumitomo Chemical Co., Ltd.
Chuo-ku, Tokyo, Japan

Kentaro Tanaka
Research Director
Shionogi & Co., Ltd.
Shionogi Research Laboratories
Japan Health Sciences Foundation
Chuo-ku, Tokyo, Japan

Nobuo Uemura
Economic Division
Ministry of Health & Welfare
Chiyoda-Ku, Tokyo, Japan

J. de Flines (BIAC)
President, Programme Committee
on Industrial Biotechnology of
the Netherlands
Wassenaar, Netherlands

Steinar Pedersen
Director of Research
Norgen A/S, Oslo
Norway

Svetlana Broman
Senior Project Manager
Industrial Development Department
National Industrial Board
Stockholm, Sweden

Peter Lange
BMFT
Bonn, Germany

Orio Ciferri
Scientific Attaché
Embassy of Italy
Ottawa, Ontario
Canada

Gabriel Milanesi
Instituto di Genetica Biochimica
ed Evoluzionistica del C.N.R.
Pavia
Italy

Tetsuo Iino
Dept. of Genetics, Faculty of Science
Tokyo University, Bunkyo-ku
Tokyo, Japan

Yoshinori Takashima
Sumitomo Chemical America Inc.
New York, USA

Shinya Tsuru
Director, Research Planning and
Liaison Officier
National Institute of
Agro-Environmental Sciences
Ministry of Agriculture, Forestry
and Fisheries
Tsukuba-gun, Ibaragi-ken
Japan

Akio Yamamoto
Office of Biotechnology
Ministry of Agriculture, Forestry
and Fisheries
Chiyoda-ku, Tokyo, Japan

M.C.F. van den Bosch
Project Leader for Biotechnology
Directorate-General for Industry and
Regional Policy
The Hague, Netherlands

Manuel Carrondo
Lab. Biochemical Engineering
Faculdade de Ciencias e Tenologia
Univ. Nova Lisboa
Portugal

Kerstin Eliasson
Head of Section
Ministry of Education
Stockholm, Sweden

Asalie Larsson
Head of Section
Ministry of Industry
Stockholm, Sweden

Ragnar Ohlson (BIAC)
Vice-President R&D
AB Karlshamns Oljefabriker
Karlshamm, Sweden

Paul Peringer
Ecole Polytechnique fédéral, Lausanne
Lausanne, Switzerland

Ronald F. Coleman
Government Chemist
Laboratory of the Government Chemist
Department of Trade and Industry
London, United Kingdom

Scott Baker
Special Assistant to the
Assistant Administrator for
Research and Development
US Environmental Protection Agency
Washington, D.C.
USA

James W. Falco
Director, Office of Environmental
Processes and Effects Research
Us Environmental Protection Agency
Washington, D.C.
USA

I.L. Pep Fuller
Director
Chemical and Advanced Technology Trade
Policy
Office of the US Trade Representative
Executive Office of the President
Washington, D.C.
USA

Leonard J. Guarraia (BIAC)
Director, Science and Technology Policy
03G Monsanto Company
St. Louis, Missouri
USA

David Kingsbury
Assistant Director
Biological, Behavioral
and Social Sciences
National Science Foundation
Washington, D.C.
USA

Charlotte af Malmborg
Swedish National Board for Technical
Development
Stockholm, Sweden

Karl Heusler (BIAC)
Director of Research
CIBA-GEIGY AG
Basle, Switzerland

Nicolas Roulet
Office fédéral de l'éducation et de
la science
Berne, Switzerland

Brian M. Richards (BIAC)
Chairman
British Biotechnology
Cowley, United Kingdom

John Cohrssen
Attorney Advisor
Council on Environmental Quality
Washington, D.C.
USA

Robert D. Fluss (BIAC)
Director
Thier World and International
Organizations
Merck and Co., Inc.
Rahwy, New Jersey
USA

James Glosser
USDA/APHIS
Washington, D.C.
USA

Daniel Jones
USDA Office of Agricultural Biotechnology
Washington, D.C.
USA

Edgar L. Kendrick
USDA/Science and Education
Washington, D.C.
USA

Henry Miller
Special Assistant to the
Commissioner
FDA
Rockville, MD
USA

Elizabeth Milewski
Special Assistant for Biotechnology
to the Assistant Administrator for
Pesticides and Toxic Substances
US Environmental Protection Agency
Washington, D.C.
USA

Lynn Myhal
Special Assistant to the
Deputy Assistant Secretary
for Basic Industry
US Department of Commerce
Washington, D.C.
USA

Ralph Ross
USDA/ARS
Washington, D.C.
USA

Vladimir Glisin
Direktor
Centraza Genetski Inzenjering
Belgrad, Yugoslavia

Vladimir Maric
Profesor Prehranbeno-Biotechnoloskog
Fakulteta
Zagreb, Yugoslavia

OECD Secretariat

Science and Technology Policy Division, DSTI
Mr. John Bell
Mr. Salomon Wald
Miss Bruna Teso
Miss Nancy Field (Consultant)
Mr. Rüdiger Hören (Consultant)

Chemicals Division, Environment Directorate
Ms. Fran Schulberg

WHERE TO OBTAIN OECD PUBLICATIONS
OÙ OBTENIR LES PUBLICATIONS DE L'OCDE

ARGENTINA - ARGENTINE
Carlos Hirsch S.R.L.,
Florida 165, 4º Piso,
(Galeria Guemes) 1333 Buenos Aires
Tel. 33.1787.2391 y 30.7122

AUSTRALIA - AUSTRALIE
D.A. Book (Aust.) Pty. Ltd.
11-13 Station Street (P.O. Box 163)
Mitcham, Vic. 3132   Tel. (03) 873 4411

AUSTRIA - AUTRICHE
OECD Publications and Information Centre,
4 Simrockstrasse,
5300 Bonn (Germany)   Tel. (0228) 21.60.45
Gerold & Co., Graben 31, Wien 1 Tel. 52.22.35

BELGIUM - BELGIQUE
Jean de Lannoy,
avenue du Roi 202
B-1060 Bruxelles   Tel. (02) 538.51.69

CANADA
Renouf Publishing Company Ltd/
Éditions Renouf Ltée,
1294 Algoma Road, Ottawa, Ont. K1B 3W8
Tel. (613) 741-4333
Toll Free/Sans Frais:
Ontario, Quebec, Maritimes:
1-800-267-1805
Western Canada, Newfoundland:
1-800-267-1826
Stores/Magasins:
61 rue Sparks St., Ottawa, Ont. K1P 5A6
Tel. (613) 238-8985
211 rue Yonge St., Toronto, Ont. M5B 1M4
Tel. (416) 363-3171
Federal Publications Inc.,
301-303 King St. W.,
Toronto, Ontario M5V 1J5
Tel. (416)581-1552

DENMARK - DANEMARK
Munksgaard Export and Subscription Service
35, Nørre Søgade, DK-1370 København K
Tel. +45.1.12.85.70

FINLAND - FINLANDE
Akateeminen Kirjakauppa,
Keskuskatu 1, 00100 Helsinki 10   Tel. 0.12141

FRANCE
OCDE/OECD
Mail Orders/Commandes par correspondance :
2, rue André-Pascal,
75775 Paris Cedex 16
Tel. (1) 45.24.82.00
Bookshop/Librairie : 33, rue Octave-Feuillet
75016 Paris
Tel. (1) 45.24.81.67 or/ou (1) 45.24.81.81
Librairie de l'Université,
12a, rue Nazareth,
13602 Aix-en-Provence   Tel. 42.26.18.08

GERMANY - ALLEMAGNE
OECD Publications and Information Centre,
4 Simrockstrasse,
5300 Bonn   Tel. (0228) 21.60.45

GREECE - GRÈCE
Librairie Kauffmann,
28, rue du Stade, 105 64 Athens Tel. 322.21.60

HONG KONG
Government Information Services,
Publications (Sales) Office,
Information Services Department
No. 1, Battery Path, Central

ICELAND - ISLANDE
Snæbjörn Jónsson & Co., h.f.,
Hafnarstræti 4 & 9,
P.O.B. 1131 – Reykjavik
Tel. 13133/14281/11936

INDIA - INDE
Oxford Book and Stationery Co.,
Scindia House, New Delhi 110001
Tel. 331.5896/5308
17 Park St., Calcutta 700016   Tel. 240832

INDONESIA - INDONÉSIE
Pdii-Lipi, P.O. Box 3065/JKT.Jakarta
Tel. 583467

IRELAND - IRLANDE
TDC Publishers - Library Suppliers,
12 North Frederick Street, Dublin 1
Tel. 744835-749677

ITALY - ITALIE
Libreria Commissionaria Sansoni,
Via Lamarmora 45, 50121 Firenze
Tel. 579751/584468
Via Bartolini 29, 20155 Milano   Tel. 365083
Editrice e Libreria Herder,
Piazza Montecitorio 120, 00186 Roma
Tel. 6794628
Libreria Hœpli,
Via Hœpli 5, 20121 Milano   Tel. 865446
Libreria Scientifica
Dott. Lucio de Biasio "Aeiou"
Via Meravigli 16, 20123 Milano   Tel. 807679
Libreria Lattes,
Via Garibaldi 3, 10122 Torino   Tel. 519274
La diffusione delle edizioni OCSE è inoltre
assicurata dalle migliori librerie nelle città più
importanti.

JAPAN - JAPON
OECD Publications and Information Centre,
Landic Akasaka Bldg., 2-3-4 Akasaka,
Minato-ku, Tokyo 107   Tel. 586.2016

KOREA - CORÉE
Kyobo Book Centre Co. Ltd.
P.O.Box: Kwang Hwa Moon 1658,
Seoul   Tel. (REP) 730.78.91

LEBANON - LIBAN
Documenta Scientifica/Redico,
Edison Building, Bliss St.,
P.O.B. 5641, Beirut   Tel. 354429-344425

MALAYSIA/SINGAPORE -
MALAISIE/SINGAPOUR
University of Malaya Co-operative Bookshop
Ltd.,
7 Lrg 51A/227A, Petaling Jaya
Malaysia   Tel. 7565000/7565425
Information Publications Pte Ltd
Pei-Fu Industrial Building,
24 New Industrial Road No. 02-06
Singapore 1953   Tel. 2831786, 2831798

NETHERLANDS - PAYS-BAS
SDU Uitgeverij
Christoffel Plantijnstraat 2
Postbus 20014
2500 EA's-Gravenhage   Tel. 070-789911
Voor bestellingen: Tel. 070-789880

NEW ZEALAND - NOUVELLE-ZÉLANDE
Government Printing Office Bookshops:
Auckland: Retail Bookshop, 25 Rutland Stseet,
Mail Orders, 85 Beach Road
Private Bag C.P.O.
Hamilton: Retail: Ward Street,
Mail Orders, P.O. Box 857
Wellington: Retail, Mulgrave Street, (Head
Office)
Cubacade World Trade Centre,
Mail Orders, Private Bag
Christchurch: Retail, 159 Hereford Street,
Mail Orders, Private Bag
Dunedin: Retail, Princes Street,
Mail Orders, P.O. Box 1104

NORWAY - NORVÈGE
Tanum-Karl Johan
Karl Johans gate 43, Oslo 1
PB 1177 Sentrum, 0107 Oslo 1Tel. (02) 42.93.10

PAKISTAN
Mirza Book Agency
65 Shahrah Quaid-E-Azam, Lahore 3 Tel. 66839

PHILIPPINES
I.J. Sagun Enterprises, Inc.
P.O. Box 4322 CPO Manila
Tel. 695-1946, 922-9495

PORTUGAL
Livraria Portugal,
Rua do Carmo 70-74, 1117 Lisboa Codex
Tel. 360582/3

SINGAPORE/MALAYSIA -
SINGAPOUR/MALAISIE
See "Malaysia/Singapor". Voir
« Malaisie/Singapour »

SPAIN - ESPAGNE
Mundi-Prensa Libros, S.A.,
Castelló 37, Apartado 1223, Madrid-28001
Tel. 431.33.99
Libreria Bosch, Ronda Universidad 11,
Barcelona 7   Tel. 317.53.08/317.53.58

SWEDEN - SUÈDE
AB CE Fritzes Kungl. Hovbokhandel,
Box 16356, S 103 27 STH,
Regeringsgatan 12,
DS Stockholm   Tel. (08) 23.89.00
Subscription Agency/Abonnements:
Wennergren-Williams AB,
Box 30004, S104 25 Stockholm Tel. (08)54.12.00

SWITZERLAND - SUISSE
OECD Publications and Information Centre,
4 Simrockstrasse,
5300 Bonn (Germany)   Tel. (0228) 21.60.45
Librairie Payot,
6 rue Grenus, 1211 Genève 11
Tel. (022) 31.89.50
United Nations Bookshop/
Librairie des Nations-Unies
Palais des Nations,
1211 – Geneva 10
Tel. 022-34-60-11 (ext. 48 72)

TAIWAN - FORMOSE
Good Faith Worldwide Int'l Co., Ltd.
9th floor, No. 118, Sec.2
Chung Hsiao E. Road
Taipei   Tel. 391.7396/391.7397

THAILAND - THAILANDE
Suksit Siam Co., Ltd.,
1715 Rama IV Rd.,
Samyam Bangkok 5   Tel. 2511630

INDEX Book Promotion & Service Ltd.
59/6 Soi Lang Suan, Ploenchit Road
Patjumamwan, Bangkok 10500
Tel. 250-1919, 252-1066

TURKEY - TURQUIE
Kültur Yayinlari Is-Türk Ltd. Sti.
Atatürk Bulvari No: 191/Kat. 21
Kavaklidere/Ankara   Tel. 25.07.60
Dolmabahce Cad. No: 29
Besiktas/Istanbul   Tel. 160.71.88

UNITED KINGDOM - ROYAUME-UNI
H.M. Stationery Office,
Postal orders only: (01)211-5656
P.O.B. 276, London SW8 5DT
Telephone orders: (01) 622.3316, or
Personal callers:
49 High Holborn, London WC1V 6HB
Branches at: Belfast, Birmingham,
Bristol, Edinburgh, Manchester

UNITED STATES - ÉTATS-UNIS
OECD Publications and Information Centre,
2001 L Street, N.W., Suite 700,
Washington, D.C. 20036 - 4095
Tel. (202) 785.6323

VENEZUELA
Libreria del Este,
Avda F. Miranda 52, Aptdo. 60337,
Edificio Galipan, Caracas 106
Tel. 951.17.05/951.23.07/951.12.97

YUGOSLAVIA - YOUGOSLAVIE
Jugoslovenska Knjiga, Knez Mihajlova 2,
P.O.B. 36, Beograd   Tel. 621.992

Orders and inquiries from countries where
Distributors have not yet been appointed should be
sent to:
OECD, Publications Service, 2, rue André-Pascal,
75775 PARIS CEDEX 16.

Les commandes provenant de pays où l'OCDE n'a
pas encore désigné de distributeur peuvent être
adressées à :
OCDE, Service des Publications. 2, rue André-
Pascal, 75775 PARIS CEDEX 16.

71602-03-1988

OECD PUBLICATIONS, 2, rue André-Pascal, 75775 PARIS CEDEX 16 - No. 44341 1988
PRINTED IN FRANCE
(93 88 04 1) ISBN 92-64-13072-1

7641